for climate action that recognizes these realities, and which is fierce, funny, open-hearted and never prescriptive."

—**Meehan Crist**, writer in residence in Biological Sciences,
Columbia University

"*The Conceivable Future* is a manifesto for parenting with purpose in the era of climate upheaval."

—**Michael Levitin,** author of *Generation Occupy*

THE CONCEIVABLE FUTURE

THE CONCEIVABLE FUTURE

Planning Families and Taking Action in the Age of Climate Change

MEGHAN ELIZABETH KALLMAN AND

JOSEPHINE FERORELLI

FOREWORD BY ELIZABETH RUSH

ROWMAN & LITTLEFIELD
Lanham • Boulder • New York • London

Published by Rowman & Littlefield
An imprint of The Rowman & Littlefield Publishing Group, Inc.
4501 Forbes Boulevard, Suite 200, Lanham, Maryland 20706
www.rowman.com

86-90 Paul Street, London EC2A 4NE

British Library Cataloguing in Publication Information Available

Library of Congress Cataloging-in-Publication Data
Names: Kallman, Meghan Elizabeth, author. | Ferorelli, Josephine, author.
   Title: The conceivable future : planning families and taking action in the
      age of climate change / Meghan Elizabeth Kallman and Josephine
      Ferorelli.
   Description: Lanham : The Rowman & Littlefield, [2023] | Includes
      bibliographical references and index. | Summary: "The impact of climate
      change has created a global concern about the future of our reproductive
      health. The Conceivable Future creates a conversation of what family
      planning in the era of climate consequences is while being a stimulating
      guide to a balanced life of participating within climate activism"--
      Provided by publisher.
   Identifiers: LCCN 2023023952 (print) | LCCN 2023023953 (ebook) | ISBN
      9781538179697 (cloth ; alk. paper) | ISBN 9781538179703 (epub)
   Subjects: LCSH: Family planning. | Childfree choice. | Climatic
      changes--Effect of human beings on. | Environmentalism.
   Classification: LCC HQ766 .K285 2023 (print) | LCC HQ766 (ebook) | DDC
      363.9/6--dc23/eng/20230727
   LC record available at https://lccn.loc.gov/2023023952
   LC ebook record available at https://lccn.loc.gov/2023023953

Dedication

With our most active hope, to all the children born in the year 2100.

With love and gratitude to our mothers, Beth Frost and Martha Saxton.

And to Jamie Topper, who makes music out of dark matter.

Contents

Acknowledgments

We owe a great deal to many people who have supported this project at different phases over the past decade. To our early organizing colleagues who resonated with this conversation, we thank you.

We also thank those who generously and vulnerably gave of their stories: in addition to the many people who have participated in Conceivable Future house parties over the years, we thank Marla Marcum, Sandie Ha, Adelle Monteblanco, Jade Sasser, Chris Begley, Jen Mendoza, Jayeesha Dutta, Yudith Nieto, Juliste Gogolinski, Juan-Pablo Velez, Andy Posner, Dan Rosenberg, Daisy Bassen, Lex Rofeberg, Jay O'Hara, Mariama White-Hammond, Chris Davies, Eli Pushkarewicz, Matthew Goldberg, Shoshana Friedman, Susie Burke, the RI-based Polycule (anonymous), Rick Lent, Roger Luckmann, Tina Grosowsky, Arnie Epstein, Tilman Gerald, Christopher Bradshaw, and Jennifer Boylan.

We are grateful for the strategic and organizational support of many people, including our editor Suzanne Staszak-Silva, our agent Susan Golomb, Mariah Stovall, Madeline Ticknor, Joanna Wattenberg, Elizabeth Rush, Cristina Moon, Brian Ashby, Brett Story, Victoria Lean, Simone Zucker, Britt Wray, Camila Thorndike, Christy LeMaster, Alyce Henson, Christa Donner, Kelly Nichols, Olga Bautista, Devin Malone, Elise Zelichowski, Sylvia Hood Washington, Elaine Enarson, Marlene Gerber Fried, Amara Miller, Heather McMullen, Jonathan Meiburg, Miranda

Massie, Beth Doglio, Michael Levitin, Cherri Foytlin, Eleanor Ray, Steve Albini, Merritt Juliano, Matthew Schneider-Mayerson, Rebecca Zorach, Spencer Merolla, Kate Schapira, and Laura Mandelberg. Josephine's fellow residents at The Ragdale Artist Community listened graciously to early wisps of this book. We are grateful for that opportunity, and encouragement. Jesca Hoop and Frazey Ford made albums that became our unofficial soundtrack: we thank you.

Many people generously gave their time and insight to drafts of this book, and to you we are deeply indebted: Elizabeth Rush, Eiren Caffall, Gretchen Lida, Cathy Lantz, Mei Liou, David Holzman, Kristi Del Vecchio, Riley Stevenson, Beth Fussell, Thalia Viveros-Uehara, Beth Frost, and Martha Saxton.

In order to tell this story we each had to draw deeply from all our relationships, both because writing is a marathon, and because our subject is a life worth living.

Meghan extends special gratitude for the love, support, mischief and inspiration of those who make life so beautiful: Jay O'Hara, Zoë Gardener, Jessica Brown, Kassie Stovall, Morgan Lord, Meg Klepack, Lisa Pellegrino, Michael Gore, Anna Kallman, Jamie Pahigian, Soda Mamy Lo, Alison Nihart, Mira McLennan, Wendy Rowland Reynoso, Dana Borelli-Murray, Nathan Ackroyd, Adam Westbrook, Mollie Kostka, Ora Szekely, Stephanie Malin, Natalie and Joanne Robelin, Liza Mindemann, Matt Cameron, Russell Mott, Rachel Gordon, and the entire raucous next generation(s) of the Kallman/Noyes/Irwin/Dechert/Robelin families. My students always push me to think harder and more expansively, and I am indebted to many cycles of them by now. Those who came before me, including Denny and Charlotte Frost, Ray Kallman, Nancy Rudolph, and Kate Mott, have been lights to see by. To my political and activist colleagues—you know who you are—thanks for being in it with me, and for your sincerity in the face of everything. To Josephine, whose friendship, humor, and perspective are unmatched, and whose taste in music is also unmatched. To ERB, always.

Josephine thanks the following beloved people: Rebecca Lyon, Hope Dinsmore, Justin Dennis, Jeff Milam, Katharine Kharas, Susannah Ribstein, Anne Goldberg, Emily Winter, Tim Merrill, Jem Dillon, Lauren Beitler, Corrigan Nadon-Nichols, Mei Liou, Angela Watkins, the Ruiners and the Adapt or Die Social Club, Chloe Holzman, David Holzman, Dan Reinhard, Isabel Slokar, Zoë Spring, Francesca Giordano, and my Teachers with a capital T: Leslie Riley, Gabriel Halpern, Rich Logan, and Lani Granum.

With special thanks to Linda and Dennis Salveter, who welcome me; to my dear sisters Stephanie Salveter, Erin Salveter, and Frances Perkins; my beloved brother, Francesco Saxton, and to my sweet nibling, who is the same age as this book, Val Saxton. There aren't words for the depth of gratitude I feel for the loving memory of Martha Saxton and Enrico Ferorelli, but these 100,000-ish are an attempt. Depending on how you count, this manuscript took a frantic six months or an extraordinary ten years to write. In either case, Meghan has been a lifeline, an inspiration, and a singular friend. And finally, Christopher Salveter came to all our noise shows. He has made this book, this decade, and this view of the future, not just possible but a joy.

Foreword

SOMETIME AROUND 2018, I STARTED THINKING ABOUT HAVING A baby, in earnest. I was in a loving relationship with a person I could imagine parenting alongside. I'd just published a book. I had a job and a community. (Not that you *need* any of these things to parent, except, of course, the latter.)

Finally, after many years of waiting, the time felt right.

Except, it also felt all wrong.

I'd spent much of the previous decade listening to the stories of those people who live alongside our rapidly changing coastline, had, in fact, written the aforementioned book about (and with) them, chronicling the early impacts that climate change was having on low-lying communities all around the country. I knew in my body, mind, and spirit that should I choose to birth a baby into the world, the world I would welcome them into would be one in a profound and unsettling state of flux. Nothing would be guaranteed.

I tried to hold my desire to parent alongside what I knew of the climate crisis. As a white, upper-middle-class woman in the United States, I knew that my day-to-day existence—eating, working, even potentially parenting—breaks down the web of life upon which we all depend. I also knew that my hoped-for baby would live in a warmer world than today's; one I had been conditioned to imagine as more volatile, less secure. I was ashamed to share my want with other climate people; ashamed to talk with my husband about how often the idea of climate change and an imperiled future arose when I started to think about conception. I felt like a buzzkill and like a selfish child who wanted what she wanted no matter the cost.

I sat at this strange crossroads for a while.

The decision felt like an impossible one to make (the authors of this book call this quandary the Impossible Question). Was I choosing between parenting and not parenting, mothering and not mothering? Those binaries, I knew, somehow missed the point. But I still wasn't sure how to proceed nor what the point was exactly. (Oh, how I wish a book like the one you hold now in your hands existed back then.)

What happens next is difficult to describe, in part, because nothing happened. I continued to teach my classes, to cook supper and take out the trash, continued to write and bike and walk and unclog the shower drain. I even went on a months-long reporting trip to the edge of a disintegrating glacier in Antarctica. And through all that, amid all that, the deep desire to make of my body a home for another, to accompany that being through this life— on this awesome and broken planet—that desire never wavered. I watched. I waited. But the shift I expected didn't arrive.

Sometime during that strange, elongated season, I discovered the Conceivable Future website, which gathered together the testimonies of dozens of other people who, like me, were trying to work out the ways in which climate change was impacting one of the most personal and fundamental aspects of our lives. As I scrolled through video testimonies, I encountered strangers talking about how their deepest desires intersected with what they most fear—and what they had chosen to do as a result. The pregnant professor explained that, for her, having a child was a twofold commitment: first, to living her life as meaningfully as possible, despite her cynicism at the current state of things; second, to the world her child will inherit, which has spurred her to participate in the fight for climate justice and workers' rights. The Puerto Rican student of conservation biology vowed to raise his future children to be both critical thinkers and caretakers of the earth, while acknowledging that being able to make this choice is a privilege not equally afforded to all. The young man who grew up next to a coal-fired power plant planned to get a vasectomy because of the physical burden he carries in his body and the ecological grief that weighs on his heart. Despite how unique each voice was, there was also an uncanny familiarity to it all, a sense of eavesdropping not only on their thoughts but my own.

I wasn't the only one who struggled with how to proceed.

The heaviness that comes from facing something difficult alone lifted a little.

In addition to all the staccato starting and stopping of speech, the *um*s and eyeblinks, the searching for language to describe the indescribable, what united each monologue was the speaker's attempt to regain control in a world where individuals so easily feel powerless. I was moved by the moment when, in the middle of Meghan's video, she shifts her gaze toward the corner of the room and tries to find a way to talk about her niece being born on a day when the temperature reached 105 degrees in northern Vermont—how she was born onto an Earth hotter and less stable than any humans have ever before inhabited. Meghan's gaze lingers on the dust motes, because what do you do with that knowledge? How can we act when the things we depend upon have become undependable?

This book is oriented toward that question: sitting with the difficulty of it and then exploring what you can do once you commit to living in this world fully, and all that it asks of us. *The Conceivable Future* won't give you answers—only you can choose how you want to proceed—but it will give you different ways of thinking about what is possible and how to work toward a livable future for all, today. It will also offer you a broader context in which you can understand the multiple treats to reproductive justice that have occurred throughout history, and how the moment we currently occupy both is and isn't new. In this sense, *The Conceivable Future* is also about building community across time and space.

The more I read, the more I was reminded that no one sits alone at this crossroads. (Though the fossil-fuel executives would like us to believe otherwise.)

With time I realized I didn't have to choose between becoming a mother or not becoming a mother. There was, in fact, no this or that, no binary at all. What I wanted was a world in which life (human and nonhuman life) was always possible, not mandatory, just possible. I wanted a world in which a person always has the right to decide what happens in their own body, free from the coercion of corporate greed or fear of firestorms and unclean air. When I witnessed the real scale of my want, I saw that no single decision I made could deliver that future. Instead, I would have to get together with other people to bring that world into being. Perhaps this book's greatest gift is that it presents dozens of easy inroads to community organizing; by illustrating how all of us are already members of multiple communities, it suggests we need only to recognize them as such in order to begin to leverage our people power to produce meaningful change.

It doesn't matter whether or not I gave birth.

What I choose was to labor.

I chose to labor and labor and labor.

For only with our sweat and sweet kinetics will we make another world possible.

Elizabeth Rush
Providence, Rhode Island, 2023

Introduction

"Our Planet Is So Fucked that Some Women are Choosing to Not Have Kids," one headline blazes. "Climate Change Is So Scary, it's Making These Women Reconsider Having Children," another screams. The hyperventilation goes on and on: "The End of Babies," "To Breed or Not to Breed," "A World Without Children."

The featured image, too, is one we've seen in all its permutations: the torso of a slender woman—often white—her hand resting protectively on her pregnant belly. The roundness of the belly is superimposed with an image of the earth. The background is a dry wasteland. Pregnant belly, parched earth, blue marble planet: this is the aesthetic shorthand that our predicament is reduced to. The predicament *itself* often emerges as a question: "Can I have a child in the age of climate change?" How do we imagine the future, think about parenting, about love and families and political action, in a time of worsening environmental crisis?

This predicament is why you're here.

We thank you for being here.

We acknowledge the journey that brings you here.

To limit global temperature rise to a level that can still sustain human life, we have huge changes to put in motion before 2030. As we write this, that's seven years to make the most radical transformation to our societies and economies that humans have ever attempted. For many people—including us, your authors—it's also the heart of adulthood, and the balance of our reproductive years. Confronted with such an immense challenge, such terrifying prospects, and political torpor where there should be action instead, it's natural to want to turn away in fear and despair. A proper headline might read "The Climate Crisis Is Harming Our Families and Threatening Our Reproductive Sovereignty." Those of us in this predicament have a collective responsibility: our work is to define and defend what's best in human nature and culture, and honor the world in which we live. This is why we get up every day—to protect all that is beautiful, beloved, and vulnerable, despite worsening odds.

Young adults have a unique experience of this crisis: as we work to make sense of, and mitigate, this crisis, we are also making many of the biggest decisions of our lives. As of 2023, there are approximately 1.8 billion millennials and two billion Gen Z-ers, worldwide. Chief among our decisions over the next decade are whether to have children, and how to parent the children we have. Such decisions have always required that we imagine a future for those children. That imaginative exercise, while not universal, is central to the human experience. Today, that future is volatile, dangerous. Across the world, young people are asking themselves: Can I have a child, or another child? How do I protect my children? How do I care for a family under growing threats?

Some of these questions are culturally specific, some arise from the acute pressures of a particular time. But if you're reading this, you're likely already aware that this moment in history presents some of the most challenging conditions in which to live these consequential years of your life. Whether you're moved to become a parent yourself or not, whether you are already a parent, wanting to be a parent, or you're struggling to decide, the fact that you have this book in your hands means you've grown up and grown older to warning bells sounding louder each year.

We know that the climate crisis is an emergency, but many of us experience a disconnect between what we know (it's serious), what we feel (an ambient sense of dread), and what we can do about it (we wish we knew). If your experience has been anything like ours, it's probably been weird. So much of the rhetoric around the climate crisis says we must avert disaster "for the children," and when we love our children, our students, our siblings and niblings[1] and young friends, we understand this viscerally. But in an equally true way, we *are* the children. We were born into this crisis, and we have only to look at our parents to know what it is to act—and not—for the children. With this awareness, that we are knotted to the past and future by family, it may be easier to shift from fear, blame, and guilt toward solutions.

WHO WE ARE

We are Meghan Kallman and Josephine Ferorelli, cofounders and codirectors of Conceivable Future, an organization we created in 2014 to start a national conversation about the climate crisis and its effects on all our reproductive lives.

We met one night when we both went to see our mutual friend Jamie Topper perform music she'd composed, on instruments she'd built, to accompany a physicist's lecture on dark matter. She introduced us to each other before her concert began, saying "you're both climate people, you should talk." We were both activists, feeling alienated and discouraged with the climate movement. And we were both thirty years old at the time, so family planning decisions were starting to feel both more pressing and more impossible.

Our first conversation, whispered in a dark auditorium, was about these fears, disappointments, and longings. And as we felt an instant spark of friendship, we recognized the power of sharing feelings as key to our organizing model. Since then, we've been bringing people together at house parties across the country for conversations about the reproductive crisis that is climate change. We built a website, began recording people's testimonies, and helped groups organize their own events. We found that people badly needed to talk. And since then, research has emerged proving that these concerns are global: we've just facilitated part of the conversation.

Our participants share their stories with each other, and, if they choose, with the wider world through video testimonies. Amid a proliferation of numbers, graphs and thousand-page committee reports, the lived, spoken truth has the power to cut through the noise. In nearly a decade of house parties, we've been educated, shocked, moved to tears, and pushed forward by displays of personal courage. Through these conversations—about reproduction and parenting and family—we've learned about people's fears, hopes, and lives. It's taught us as much about the present as the future. And through this project our trust, friendship, and shared understanding has deepened.

But each of us—Meghan and Josephine—came to this work following a different trajectory.

Meghan was born in rural New Hampshire, to parents who were a biology teacher and a contractor. Her family is big and rambunctious and multigenerational, made of a dizzying mix of biological and step- and adopted siblings, parents, stepparents, grandparents. They were hardscrabble and lived close to the ground—heating with wood, preserving food from the garden— and watched, even in the 1980s, as the climate began to shift around them. Having formative experiences of poverty, including housing insecurity, left her salty and mischievous and joyful and ambitious: simultaneously in love with the world, curious, mad about the injustice, and determined to prove herself in the face of it. Meghan went on to get a PhD and is now a professor at a public research university in Massachusetts and a state senator in Rhode Island, representing her beloved city of Pawtucket, as well as Providence. From this tangle of contradictions she learned to organize, drawn into the protests against the US invasion of Iraq in 2002 and making her way into climate activism as a young adult.

Josephine was raised in the wholesale fur and flower district of New York City by an American history professor mom and an Italian photojournalist dad. She was shy, bookish, the opposite of athletic, and in 1992, at nine years old, she read an Archie comic about the recently popularized concept of global warming, and developed a life-altering horror of a hotter world. She and her older brother Francesco attended a prep school, receiving a top-notch academic education and a distressing social one. The fast and ambitious culture she grew up in was never a comfortable fit. She left for the University of Chicago in a packed station wagon with her parents on September 12, 2001, a cloud of smoke over lower Manhattan receding in the rearview mirror. She found home in gray, ample, inland Chicago, and stayed. In her twenties she attacked her carbon footprint through urban agriculture, biking, long-distance rail, thrifting or dumpstering everything. She wrote dystopian graphic novels and a not-very-funny comic strip called *The Lighter Side of Climate Change* while waiting tables at a vegan restaurant, selling stationery, and ultimately learning to teach yoga. Cultivating a body practice, spending time in the Occupy encampment, and a few romantic and family events marked her shift from inward focus to outward focus and activism.

Everyone's history—including both of ours—is a part of how we live and feel and think. Over time, we have developed some strong habits of partnered organizing that reflect these histories, and our values. We are a two-person political unit at Conceivable Future. We're longtime collaborators and fast friends. The nature of our commitments to each other and to the world are rooted in our identities as women, and in our vision of the world we want to see. We are committed to feminism and equity. We're intentional about our collaborative process. Every time we speak with a reporter, for instance, we make sure that both of us are present, so that we can talk *with* each other and never *for* each other. We share credit and responsibility. We prioritize our friendship.

In this book, we express our shared values, which we know are not universal. And while we endeavor to be transparent and truth-telling in all that we do, we have tended not to disclose our own decisions, or even our own thought processes about our own family lives. At the end of nearly every conversation with a reporter, we are asked about our own decisions—do we have children? Will we? We demur.

We do this because it's the *fact of the question*, and not our own answers, that matter in this work. We aim to crack open a space with less judgment and more compassion around issues of family, parenting, and political action. So, we don't talk about our own reproductive lives because in a 1,200-word article there's almost no way to do so without sounding prescriptive, like our own decisions or life paths are the ones we're advocating for everybody. Our own circumstances are not more or less right than anyone else's. And after a decade of doing our work while living our lives, we both have a much deeper appreciation of the role of chance in every life's contours. If we announced publicly what either of us were thinking or doing at any given time (which has changed so much for each of us over the past decade), those statements would likely have collapsed that open space we've created.

We're all here because we know we've got a problem
The animating idea of Conceivable Future—that environmental collapse is now a major factor in family planning—is more alarmingly relevant than it ever was. Many groups organize around parent identities: Mothers Out Front, ClimateMama, and others, do powerful and effective climate work. But we began with a belief that, for people of reproductive age, even resolving to become a parent grows more difficult in the climate crisis.

That belief has been substantiated again and again by research. A 2019 *Business Insider* study found that a third of Americans felt that climate change should be a factor in a couple's decision about whether to have children; that number climbed to 38 percent for people between the ages of eighteen and twenty-nine.[2] And a 2021 study of youth (age sixteen to twenty-five) in ten countries found that 75 percent of respondents in all countries believe that "the future is frightening."[3] Climate and reproduction are more profoundly linked than has been acknowledged: how do you decide to have a baby as the odds of them leading a healthy, long life shrink?

As activists in the environmental movement, we'd been indoctrinated with guilt for every way we weren't conforming to the rules and regulations of Green Living. This included mixed feelings about bringing a child into the world: a child whose disposable diapers could circle the earth and whose carbon footprint would multiply our own. But beside that was a deeper well of fear: not what harm our children might do to the world, but what harm was in store for any child entering into a harsher, hotter, less just world? From this embattled place, what meaningful actions could we take that helps balance big scales of injustice?

Back when we began, feelings were not yet an acknowledged part of the climate discourse. That was baffling: most of us commit to activism because we *feel*. And these feelings emerge from a society that offers up the future as a sacrifice zone—a place determined to be inessential, which can be degraded in exchange for profit. From the beginning our goal has been the same: to bring people together to share their experiences and navigate these threats, together. And by doing this, we have aimed to transform the national climate conversation from one trapped in remote, chilly legalese to a widespread, vital, moral one.

We're writing this book to describe what we've learned through our organizing work, which we believe points at new ways to understand this predicament, and its solutions. We share our own experiences, and we talk with dozens of people who have found imaginative, effective ways to join the massive effort to right this keeling ship. Our book is an invitation to get unstuck, to find your people, to acknowledge your feelings and roll up your sleeves; to learn how to be activists and change-makers. It's as much about taking action as it is about family planning. We can't give you easy answers, but we can accompany you on this journey from paralyzing questions to transformative action for a future we imagine, together.

The current climate playbook sucks

As more voices have joined the climate and family planning conversation over the past decade, we're still finding new depth and insight. Scholars, philosophers, health professionals are contributing fresh discoveries and analyses. But clickbait articles keep coming that ignore how rising temperatures multiply injustice. Each hot take assumes a correct answer to our generation's predicament—either that hope comes in the form of a baby and to think otherwise is to worry too much, or that *not* having a baby is the best thing you can do for the environment. This cognitive dissonance grinds on us. And more importantly, it misses the point.

The point is this: a focus on the *outcome* of a personal struggle—a news profile lamenting that Laurie and Max decided not to have children because of climate worries, or denouncing Alex and Jamie who just had their sixth—hides the actual *forces that are causing everybody to struggle*. The individualist tale of correct "choices" follows the same dead-end logic that drives climate change in the first place, and buries the real political meaning of the moment. Throughout this book we refer to this bind—our generation's predicament—as the Impossible Question. We'll talk about the real dimensions of the Impossible Question in the first chapter, and share some of the stories people have told us over the years. The point of these stories is to help us all feel less alone, not to pretend there's a family-based fix for an ecological catastrophe.

By blaming—or elevating—individuals, we continue to send the message that coping with climate change is an individual responsibility, to be handled within individual households. By this logic the "correct" answer is resignation and inward focus, rather than collective political action and change-making.

But the problem has never been the kids. It is our country's intensive burning of fossil fuels. This is the reality that forces us to ask the Impossible Question.

THE CLIMATE CRISIS, UNDERSTOOD THROUGH REPRODUCTION

We live in a fossil-fuel regime that's extinguishing our futures and stifling our imaginations. It sucks. It's so entrenched it's hard to even see how bad it sucks: it shows up as increased asthma rates, superstorms, and heat waves. It's locked in through trade agreements and retirement portfolios, through cultural silence, loneliness, and guilt. Everyone on the planet lives with it in different and interlocking ways. The climate crisis is being described continually by people all over the world. Our job here is just to describe one dimension of the crisis that's under-discussed.

For us, reproduction isn't the whole story, but it reveals the heart of the story. In this sense, we think of it like a core sample: a cylindrical section of something—ice, earth, or a tree, say—that cuts through all the layers, revealing its nature and changes through time. A focus on reproduction in the context of climate change exposes the same unjust core that motivates many other social movements; it shows us what's at issue, for all of us.

As the climate changes, all stakes are raised. Some nonparents feel that their commitment to climate work would foreclose the possibility of parenting. At the same time, parents have described feeling locked out of activism, or struggling to stay involved because time and money are short, because activist spaces do not often accommodate children, or because they find themselves dismissed as "merely" a mother, not meeting the conventions of radicalism. It's clear that, despite an array of technologies with which to regulate our fertilities, no one makes reproductive "choices" freely in the face of so many economic, social, and environmental pressures. It's time to hear the full chorus of voices: people who are childless by intention or circumstance, people who are parents by intention or by circumstance, parents-to-be, and people who are undecided. Only by seeing how we are interconnected, what we share, desire, and can't bear to lose, do we build a movement.

The only "right" answer to the Impossible Question is to work together, as hard as we can, for a just and habitable future for *everyone's* children.

We believe climate action is anything that mitigates the climate crisis or its effects and that positively transforms the neoliberal capitalist regime that got us here. In this view, fighting for reproductive rights and access is climate work. Working to prevent childhood lead poisoning is climate work. Building resilient food systems is climate work. All the work that moves us toward a just, sustainable society is climate work.

Paris Accord Withdrawal and the End of *Roe v. Wade*

The same week in 2022 that we were offered a book contract, the Supreme Court in the United States did two things: it overturned *Roe v. Wade*, and it gutted the EPA's ability to enforce US climate goals. The timing of these things was consequential for this book, as it was for every other person's life on the planet. We'd mostly been thinking of Conceivable Future as climate organizing: using the lens of reproduction to organize around the climate crisis. But with *Roe* gone, suddenly we were writing a different book, doing different work, and from an utterly different vantage point: we are still accelerating toward climate collapse, but women have fewer rights even than yesterday, and are literally second-class citizens. Josephine texted Meghan on Independence Day. "Just reflecting that if we were still a British colony," she wrote, "then people could get free abortions from the NHS."

We know we're really in the shit when humor swings from the gallows like that. The point is that it's hard to imagine how profoundly these two decisions will entwine to shape the next several decades.

The *Dobbs* decision means that the idea of "choice"—reproductive and otherwise—is now legally dead. We don't have the right to choose how many children we want, or when we want to have them, and we certainly don't get to choose to live in a world where our institutions steward our shared resources responsibly, or protect our collective well-being. Forced reproduction policies are associated throughout history with fascism. *Dobbs* brought the violence of the struggle for reproductive control into plain view. Examining abortion restrictions alongside the epidemic of domestic gun deaths, Caro Claire Burke writes, "the abortion argument and the gun argument are not similar arguments. They are the same argument. The government will never get in the way of a man's desire to point at a woman and shoot. To bury his will inside of her. They took away your rights because controlling your life isn't enough, and so they want your death, too."[4] Our lawmakers authorize gendered violence when they eliminate body rights and allow arsenals to swell.

Alongside *Dobbs*, there is now, in the United States, no legal federal authority to enforce emissions reduction targets. This means, in practice, that most regulatory work will need to be done at the state and/or international levels, rather than at the federal one. The good news is that most of the work was happening there anyway. But the Paris Accord withdrawal and the attack on environmental regulations raises questions afresh about who the federal government protects, who it serves. As we lose government protections, people power is more vital than ever before.

Big calamitous moments, like the summer of 2022, cause people to seek out Conceivable Future; our inbox is a little like the doomsday clock in this regard. But it is also true that everything those Supreme Court decisions represented was already in motion. The decisions were the icing on the cake, not the cake itself.

Whether you experienced those political moments as sudden and shocking, or slouching and inexorable, we can all agree that we have our work cut out for us. Now is the best—and only—moment to exercise our collective power and imagination audaciously.

PILLARS

In the decade of organizing around these issues, including hundreds of conversations with house party attendees, activists, student groups, podcasts, reporters, friends, and family, we've developed a few principles to which we try to adhere.

Focus on the problem

The problem is not you.

The problem is that our economy is built on extraction and thrives on inequality. Our leadership has prioritized short-term profit over long-term survival. Our society elicits the worst pieces of us, rather than calling forth our better nature.

All of us are implicated in different ways, but none of us is individually, personally responsible.

So we focus on the problem. And despite the impulse to finger-wag and scold, this moment calls for tremendous compassion, as well as courage. If your adult daughter expresses hesitation about having a child because of the realities we are faced with, the appropriate answer is to listen with empathy, rather than to try to change her mind. Across the climate movement, we can let people's responses to the problem point us to the source of the problem, rather than trying to regulate those responses and/or cajole them into responding differently.

Focus on the problem.

Get together

Many climate-concerned people labor in obscurity and mourn alone. Through our house parties, the stories that we've heard are personal, painful, and overloaded with cultural baggage. We are all feeling powerful, negative emotions—whether we're able to fully acknowledge them or not. But they're often stuck inside us, where they impede—rather than inform—our action. The climate crisis is not an individual problem, although we experience it as individuals; the mental health aspects of it are affecting countless people across the world. And as psychologist Sanah Ahsan argues, we don't heal this just by looking inward. Rather, we must transform the systems that are harming us, causing our anxiety, grief, depression.[5] When we acknowledge these feelings, then we can find out what comes next.

If we don't share our feelings we often obsess privately, turning in on ourselves rather than banding together. We gatekeep ourselves and each other ("I know I drive a car, so I shouldn't talk" or "Well, are *you* a climate scientist?"), rather than identifying the bigger forces acting on our lives. The innate climate problem of scale—that we are tiny and helpless—is aggravated by isolation.

This cognitive dissonance is paralyzing and demoralizing. If you follow personal footprint reduction strategy to its logical end, the most effective action you can take for the climate alone is to kill yourself now. We don't say this to be callous or macabre: the tragic reality is that some people, no longer able to bear this devastation, have ended their lives. But the example proves the point: suicide does not solve the climate crisis, and one person fewer on the planet doesn't fix systemic injustices. We can move beyond the view that our individual consumer choices, reproductive choices, or—in the limit case, existential choices—are our most meaningful contribution to the fight against climate consequences. We are best able to do more good than harm when we join together, become bigger than ourselves, and reach for the big levers. And so we must get together.

The lightbulb was invented by candlelight. We start from where we are, in order to make the next world.

Politics is what you do

We acknowledge that the cumulative forces bearing down on us are experienced individually. Each one of us, stardust skeletons bound in flesh, has an individual experience of living, even if those experiences follow patterns. As we said before, Josephine and Meghan are different people, with different feelings and lives, who also share some similarities. As a team, we have been practicing the tricky balance of acknowledging our individual experiences, recognizing the forces that shape us, and using those experiences and knowledge to take collective action. We feel things on the scale of ourselves, but we act in concert with others to make a difference.

The default setting is: we care but don't share. We watch the world heating up, we watch our rights being stripped away, we feel helpless, tongue-tied, overwhelmed, frightened, fatigued. Our hearts may be breaking but we lapse into silence.

Which is a bummer, because a just future needs people power to create it. Suffering alone doesn't change anything.

And it might take less than we imagine to accomplish big goals. Research shows that only 3.5 percent of a population need to sustain political action for regime change.[6] That's not a small number, but it's not at all an impossible number, either. If our goal, then, is to end the fossil-fuel regime, we need to get together. Movements logically grow through relationships. Burdens are easier to bear when we share pleasure and support. And relationships grow from a shared vision of the future.

Politics is what you do collectively, not what you believe, or how you feel. We know that climate action can alleviate climate anxiety, but our action only succeeds when we're grounded in personal practices and supportive relationships. Nobody can survive on a diet of negatives. Nor can we survive in a state of simplistic optimism, suppressing fear and grief. Your politics will be animated by your beliefs, propelled by your feelings, but your *politics is what you do* to change a system, in collaboration with others.

"Believing in" climate change is not political action. Worrying to yourself about climate change is not political action. *Political action* is political action. The decision to have a kid or not may be informed by our beliefs and political orientation and identity, but it is not inherently a climate action. In other words: there is no one thing you can do in your personal reproductive life that will fix the problem of climate change.

It is no baby's fault that it was born into a carbon-intensive, violent economy. Its nonexistence wouldn't alter the systemic problem in which its parents are implicated. The only moral imperative of the Impossible Question is this: to demand the biggest systemic solutions we can imagine. Making those demands, campaigning and building better systems for everyone's children: *that* is political action.

Our vision is the conceivable future
We are describing a vision of a life worth living and a world worth living in.

Hope compels us. Hope, as Rebecca Solnit describes it, is an action rather than an expectation of a preordained future. Hope "locates itself in the premises that we don't know what will happen and that in the spaciousness of uncertainty is room to act."[7] Optimists say a tech fix is coming; pessimists say the die has been cast; both excuse themselves from action. But hope lives in the unstuck middle place, where we acknowledge that the future is unwritten. Hope is the belief that what we do matters, even though we don't know if, when, or how our actions will have an effect. What we do matters because it defines our character, our relationships, and the quality of our lives.

Climate action matters because Black Lives Matter. Climate action matters because we hold all genders to be equally precious. Climate action matters because reproductive rights are human rights, and a knowledge of history demands justice. If we hope to arrive in a just future, each step we take must be made from that material.

Solnit wrote, in her 2020 book *Recollections of My Nonexistence*, that "young women are urged to 'never stop picturing their murder.' From childhood onward, we were instructed to not do things—not go here, not work there, not go out at this hour or talk to those people or wear this dress or drink this drink or partake of adventure, independence, solitude; refraining was the only form of safety offered from the slaughter."[8] Her point was that the imaginative burden of always trying to avert the worst takes a massive psychic toll on women.

We think the same is true of the climate catastrophe—which exacts tolls across every border of inequality. In every way it's always been hard for people to dream, it's getting harder. But the point is this: until we dream of something better, the future isn't worth doing.

We elaborate this positive vision throughout the book, especially in the last chapter. But here we outline a few aspects of it that are non-negotiable. In this sense, the path is the goal.

- The conceivable future has a livable climate, in which risk and harm mitigation is well-planned and funded, and costs and benefits are equitably shared.

- The conceivable future has an economy that is regenerative and non-extractive.

- The conceivable future is just. It enshrines accessible, affordable, expansively defined and legally protected healthcare for everyone. It is anti-racist, gender-inclusive, feminist.

Reproductive justice is a guiding force of this vision because it joins together the social values of equal humanity, intergenerational commitments, racial justice, individual and collective well-being, accountability, pleasure and joy, and sexual and reproductive health.

NOTES ON INTELLECTUAL PROCESS AND LANGUAGE

We make an important distinction here around the term reproductive justice. Reproductive justice is a framework, a concept: we explore it in chapter 2, and it has mattered tremendously to our work. But reproductive justice is also a movement that emerged from a particular time and place and set of concerns. The intellectual labor of developing it has been done primarily by Black women in response to state violence and oppression intruding on their lives. While it has been an orienting framework for us, we want to be clear that we are two white women who rely on the pioneering intellectual work of the National Black Women's Health Project, Women of African Descent for Reproductive Justice, SisterSong, Black Mamas Matter Alliance, and many others. Conceivable Future is not a reproductive justice organization. Rather, it is a climate organization that uses reproductive justice concepts to address climate and environmental problems—the harms caused by the movement itself, as well as the harms the movement intends to repair—and to imagine clear, justice-based solutions. In other words, reading this book and thinking about climate through the lens of reproductive justice won't transform you into a reproductive justice activist or even ally, unless you actually join that movement: politics is what you do, not what you believe.

A word about terminology is also in order. This book moves through subjective experiences of gender, and also cultural gender norms. We strive to be reflective about that, and we strive to use language that is both welcoming and descriptive of the world that we live in. We avoid gender-deterministic language as we're able. For example, we use the phrase "pregnant people" both to be clear that people of any gender can be pregnant, and to reaffirm that women—who make up a majority of this group—are people.

We recognize that language changes and develops alongside our collective understanding, and that some of the terms we use now won't fit the world in 2040. We celebrate that evolution.

However, legacies of how gender has been understood continue to shape our lives (For instance, public health and sociological research did not use trans-inclusive language until relatively recently. That means that when we cite older research, we'll use the language that the older research used). When we talk about history, we're using historically gendered terms—we are going back in time to understand what was happening, then. Those historical terms don't map perfectly onto our categories now, but we engage with them as part of how we engage with history. We're aiming for a world that is expansive in its concept and experience of gender, and we're describing inheritances of patriarchy/hetero/cissexism, so our use of gender markers throughout the book is situational.

PART I

THE IMPOSSIBLE QUESTION

ANXIETY ABOUT A HEATING WORLD BECOMES AN INCREASINGLY pointed problem for young people as we come of age and plan families in earnest. We began organizing in 2014 to shed light on something we felt deeply, but had struggled to describe in words. As we spoke with more and more people about this murky problem, we found that it often expressed itself as a question. This question had been with many of us for as long as we could remember, though we had never heard it asked out loud:

"Can I have a child in the age of climate crisis?"

And in the absence of any sincere public discussion, this question—and the belief that there is a correct answer—was causing a private epidemic of grief, anxiety, guilt, stress, and fear. These fears have been shaking the very foundations of family planning. Through conversations we conducted across the United States we learned that people's concerns about starting families were bounded by two main questions:

- What harm will the world do to my baby?
- What harm will my baby do to the world?

In the United States, no two subjects are more polarized than climate change and reproduction: we're taught to avoid them both in polite company, certainly. And as with anything we avoid discussing, the conversations that we *do* have often lack nuance. And so we got curious about how people's understanding of the reproductive stakes of the climate crisis had already been so substantially shaped, when this was the first time many of us had ever even talked about it.

The first concern—that the world will harm babies—is well-rooted in evidence and predictions from scholars in every discipline. Whether they study the behavior of water, weather, or groups of people, experts agree that heat drives chaos: the hotter the average temperature, the more volatile our experiences become (more on all that in chapter 1).

But the second concern—that a baby will harm the world—has been shaped by decades of global policy driven by the ideological concept of "overpopulation," and fossil-funded PR campaigns directing individuals to "fix" the climate by shrinking our own carbon footprints. Many reproductive-aged people feel a burden of guilt at the prospect of creating a new human. Thus indoctrinated, we obsess over the ways we could be greener. And because the mainstream environmental movement lacks a historically grounded, justice-oriented perspective on reproduction, it's very common for primarily white, middle-class environmental activists to see population as a central driver of climate change. Media has focused largely on the white middle-class experience of these universal concerns, which pushes everybody else out of the story and makes it harder for justice issues to come into focus.

"I was a child of '90s environmentalism," says Rabbi Shoshana Meira Friedman, a Boston-area climate organizer and spiritual leader who is mother to one young child. "We were focused on individual consumption. I remember just my obsessive thinking about every time I threw something out, or every time I turned on the tap. Every time I turned on the light, every time I got in the car. I was conscious that I was using something finite that I shouldn't be using. I now catch myself when I get sucked into that kind of thinking, but for decades it was utterly pervasive in my mental model." We've found that among people considering parenthood, even non-activists report shame and ambivalence about the carbon and material waste associated with a baby.

The mainstream environmental discourse has made shaping individual behaviors its priority over the past thirty years. In doing so, it has functionally defined babies as carbon sins—while blithely ignoring the class, race, and gender problems of such a framing. Given this view, it stands to reason that anyone would have mixed feelings about bringing a child into the world. At the same time, people struggling with this ambivalence have been mocked and caricatured as fringy. Reducing one's personal carbon footprint has become a classic dead-end dilemma that obscures rather than exposes the real problem, and the real solutions.

We've been socialized to ask ourselves the Impossible Question *instead of* asking why having a baby in the industrialized world is so carbon-intensive to begin with, *instead* of taking a critical view of the systems and institutions that shape our lives. This guilt-driven individual-focus also covers up for the damage fossil fuels are already doing to babies and their families. Instead of sending the blame *up*, toward economic and political leaders, this attitude weighs *down* our generation's fundamental human right to have children. In chapters 1 and 2 we'll look more closely at the environmental, medical, and political problems we face, as well as solutions. For now, though, we're refusing to accept the Impossible Question.

Throughout many of our conversations, with reporters especially, we've been perplexed by the pervasive belief that there is a "right" answer to the question of: "Should I have a child in the age of climate crisis?" (Or more specifically: "Should *women* be having children in the age of climate crisis?" Carbon emissions associated with a child are traditionally awarded to mothers).

But the question itself is backward. When we put responsibility on individual people to make "correct" decisions about having children, we avoid naming system-level problems. It's easier to criticize people, and particularly women—for having "too many" children—than it is to call out the fossil-fuel industry for insinuating itself into every opening of our polity.

But we don't want to be the greenest individuals on a dead planet. Why have we found it easier to tell people to stop having babies than to tell corporations to stop polluting? Or to tell our legislators to stop taking fossil-fuel money?

There is no right answer to an Impossible Question. Trying to answer it "correctly" is not radicalizing; it's deadening. Fear of making a "wrong" decision can lead to paralysis, and any space to consider the real problem collapses. Accept the premise of the question and the conversation is already over: the focus and the blame are on the would-be parents—and their maybe babies—instead of the CEOs, corporate boards, and the enabling lawmakers. We are backed into a corner. If we think there's a broadly applicable "right" answer, then we've already agreed to stay in the corner, rather than challenging the forces that put us there.

It's the *fact of the question*—not any person's individual answer—that describes the moral dimensions of this crisis. Our moment requires us to live with imperfect answers, in radical uncertainty. It's understandable to want to resolve that uncertainty with a simple, correct answer. But this is not a yes-or-no question.

Instead, to help people consider their experiences of this crisis in a fuller, less judgmental, more humane way, we ask: "How is climate change shaping your reproductive life?" In an effort to hold that space, we have learned to pose this question without expectation. We ask as a way to pull back the curtain on the problem and see it in the light. When we ask "how" instead of "should," we find that although people's lives and families are varied, these concerns are widely held. And when we ask it out loud, in public, the ensuing conversation allows everyone to see the size and complexity of the problem, as well as possible paths through.

Whether or not to have kids matters deeply on a personal level: our family composition will obviously shape our personal lives, our senses of self—not to mention our sleep, our savings, and our free time. But the significance given to having a baby or not as a form of activism, as *climate action*—especially by middle-class white people—is misplaced and exaggerated. If we consider politics to be *action we take together to change systems*, then we can see that individual reproductive decisions happen *within* political systems. And for the most privileged within such systems, parenthood doesn't inherently challenge or change how the system operates.

Yes, we can make political meaning out of our personal experiences—whether we feel ambivalence, make a resolute decision, or don't follow any traditional narrative. But our individual behaviors don't approach the scale of political significance that solves problems. Let's break the habit of giving individual decisions outsize importance. At the same time, anyone who dismisses or trivializes the importance of these personal experiences is wasting our precious time and jeopardizing our future. You can tell 'em we said so. If we can make greater meaning from our own experiences, then we can take greater action. This is not the same as mistaking our own experiences *for* larger action: our actions must address the dangers we're facing, at the scale on which they occur. That means our actions must be collective.

Following this line of thought, at our house parties we begin by asking people to share why they decided to spend a perfectly good Friday evening having this conversation. We describe how Conceivable Future came to be, and we give some basic climate information. Usually we screen a couple of our recorded testimonies so people have a sense of what others have said in the past. Then we use a few different methods to consider the central question. Participants free-write, talk in small and large groups, and those who choose to can record a testimony—a short statement from the heart—that we post online and invite them to share with their own community.

It's important for us to note here that our groups are self-selecting: attendees are the kind of people who *would* spend a Friday evening talking about reproduction and climate change. They usually already know and care about both, and have the free time and sense of belonging to join this conversation. Beyond that, they are often college-educated and possess at least some of the privileges this implies. They usually have access to birth control and STI protection, the knowledge that makes those things effective, and basic family planning services. These factors shape the perspectives we're able to represent, just as much as our location in the heart of the world's carbon economy.

It's also important to note that not everyone can make the reproductive decisions that they want to, even surrounded by an array of technologies designed to help. Some people do not get to control the timing or number of their pregnancies. Others struggle to conceive, which can impose a different kind of grief on their experience. The abortion restrictions now in place throughout much of this country further constrain these options.

Having said that, by posing this open-ended question rather than pushing an agenda, we've been able to clear a little breathing room amid the competing pressures of climate, culture, family, and personal expectations. By holding this space and listening, we've learned a great deal about where people find their political strength and where they need support. What we share in the rest of this chapter comes from all over this discussion—from Conceivable Future house parties and testimonies, from conversations and interviews, from social media comment sections. These stories—which represent the range of responses we've heard to the Impossible Question—illuminate the many corners of people's hearts where these feelings live.

Deciding to Parent

For some people, having children is one of the most powerful acts of commitment that they can offer to the changing world. Some see it as motivation—love for your children becomes a reason to fight harder. "I would rather have a child and fight like hell for that child's future than give up," Katherine Fisher told us at a house party in Jamaica Plain, Massachusetts. We recorded her testimony by a west-facing window, but we'd already lost the last light of a short autumn day. A Quaker, Katherine works at a solar company and has long brown hair that falls in a wave down her back. When we first spoke, she was thirty-four years old and living in Newton, Massachusetts. When we saw her a year later, she was seven months pregnant.

Some people view having children as an opportunity to improve the world by teaching them well and shaping their values. "I can make a difference," Jahdiel Torres Cabá told us. A Latinx man, Jahdiel was thirty when he spoke with us. "I can raise someone that can develop critical thinking and be very intentional in the relationships they have in the planet with other biodiversity, and other humans as well." Jahdiel wants to parent in order to amplify what's best in himself and in human society. He views raising thoughtful and engaged children as part of his activism.

For many people—especially those who come from marginalized communities—the act of having a child is a triumph over centuries of oppression. The excellent 2022 Canadian television documentary, *The Climate Baby Dilemma*, features interviews with climate activists, including us, and concerned Canadians talking about their reproductive lives.

The filmmakers also interview queer Indigenous activist Sarain Fox, who says in the film: "My daughter will grow up on her land, and she will know her language, unlike me, who has spent all of my life yearning for that. . . . I've never met an indigenous person who didn't want to have kids because they thought the world was ending. Why would we think about not having children when we've already experienced the apocalypse?"[1] On screen she tears up. The section is emotional to watch. She continues,

> When somebody comes to your land, takes all of your resources, kills your ability to survive and have a family and then takes your children and separates them from you and tells you that everything about your existence is shameful and wrong? That's the end of the world. And that's why we can't imagine not having babies. Because if we don't have babies, the colonizers win. And that's what they want. So I'm not going to give it to them. Raising children for me is a political act, informed by history, because the choice to be a mother, to raise and protect my child, is the revolution. I think it's the only way to heal from genocide.[2]

This is a powerful framing of the climate baby dilemma. In addition to talking about her own decisions, Fox addresses several points that are often overlooked in this conversation: how context matters for whether or not something is considered "political," and how worrying about this in primarily future terms is a mark of rarely acknowledged privilege.

> My biggest fear is that we will come to a moment when I don't know if I can keep her safe. And in that place, I know why people don't want to bring precious life into this world. And so there's this real pain about imagining a future. And I think for so many populations who have had the privilege to live outside genocide? They have never had that before. Futures have been handed to them! The idea of safety, and a roof over their head, and a career, that's all normal. And so that sense of not having hope? This is new.[3]

For people born into families that never survived genocide or state violence, acknowledging this privilege can help make sense of their own experiences. Black and Jewish people considering this question in our circles—in the wake of genocides of both groups—have sometimes expressed resonant perspectives. For members of a group that has survived extermination attempts, having children can enact a commitment to the group's survival. It is, itself, a victory. But within that healing victory, Fox acknowledges the pain of a dangerous future.

Even for people who decide to become parents, climate change can massively alter the *way* that they parent. Andree Zaleska, a white forty-eight-year-old mother of two, opened her surprisingly warm carbon-negative house to a Conceivable Future house party one chilly evening. Climate awareness is an intimate part of Andree's life; she has been slowly covering herself in tattoos of extinct animals. As we recorded her testimony, she reflected that she had engaged her "kids in activism, but I've also shielded them from the worst of the climate news, from the worst of the science that I actively seek to know." Andree mourns that neither of her two sons have ever experienced a real New England winter, and are unlikely ever to do so. Nor does she expect that her own sons will have children of their own, that she will have grandchildren. By that point in time, "I just assume that will not be a good idea," she says.

Many parents express fear at what a changing climate will mean for their children's lives. "More than anything, it makes me worry about what life will be like for our daughter," one young man lamented on Facebook. And even though climate activism is often framed as "for the children," the day-to-day needs of their children are also very real for climate-concerned parents. In an already hectic life, increasingly alarming events and information may lead us to want to shut it out and let others worry about it. But even so, many search for ways to add climate action to the endless list of responsibilities that comes with raising children.

Some climate groups use parenthood—and particularly motherhood—as an organizing principle. For them, parenthood can be a lead-in to a *political* identity; a way to build collective power.

"Once you have a child," says Harriet Shugarman, the director of ClimateMama, a parents' climate organizing group, "the feeling of responsibility, coupled with the joy and the underlying fear of raising a child in the age of our climate crisis are front and center." Harriet describes how understanding the gravity, the sadness, and the horror of what likely lies ahead pushes parents to demand solutions, to seek out like-minded people, and to look to their identities as parents as a way to stay committed. "Personally," she concludes, "I often wonder if we are the 'Sarah Connors,' of the [blockbuster science-fiction] Terminator series, put here at this moment to raise our children to lead the 'rebellion'; to be the ones to fight for our future and now at all costs. Yet too few of us are actually taking the drastic steps she took."

For many parents, including the parents with whom Harriet works, the fight against fossil fuels becomes *part of their parenting responsibilities*. Climate parents often become fierce advocates for voting rights, for climate justice and for educating others on the crisis, in their effort to make the future livable for their children. Harriet thinks that the 2016 election was a huge galvanizing force for parents especially. Across political affiliations, she says, parents saw that the basic lessons of responsibility and compassion that we teach our children are not reflected by top leadership, and the outright denial of the climate crisis was a jarring refutation of reality. She witnessed this pull people with children together in important ways.

There is mounting grief among parents who do this work, but Harriet sees it as a predictable by-product of fierce love and a conviction that the future is not predetermined. "We refuse to believe our children's future is already written," she says. "We also know that we can't wait for our children to find solutions to the problems we have created. It must be us, and we are out of time."

We want to emphasize this last point, because a common piece of nonsense we've heard steadily throughout this decade is "But what if *your* baby was the one to solve the climate crisis?"

Or, as an activist in a CNN article put it more pointedly—but no less counterfactually—"Sometimes I wonder, what if Greta Thunberg's mother had not wanted to have children because of climate change?"[4] You might equally ask, "What if Rex Tillerson's mom had not wanted children?" Or the Koch brothers? The point is: we could do this all day, and it gets us nowhere.

More to the point: we need to reduce emissions 50 percent by 2030, so it would be rotten luck if all the solutions to that problem were contained in the mind of an *unborn* child. Our future children won't have the time to save us: that is *our* job, now. Still, any child born today is likely to be rightfully outraged at their Elders' procrastination. Some of those children will become activists. While we might find inspiration in Sarah Connor, we might also try living like we're the John Connors of the climate crisis, asking "What if *I'm* the person who's going to fix it all?" We don't actually have time to be the Sarah Connors anyway (barring technological advances that don't exist, like time-traveling androids and clean coal). Luckily this is not a Terminator movie, and "solving" the climate crisis is not just a task for a Chosen One. A just transition is a job that is currently hiring for *all of us.*

The tools to do it already exist.

But even with the existence of parents' climate action groups, parenthood and activism can conflict in ways that go beyond scheduling. Some forms of protest are too physically risky to participate in with children, and most parents of young children can't risk arrest because their kids depend on them. In the absence of social support like affordable childcare, and in an economy where parents often work full-time or more, children *or* activism can feel like the implicit choice. While some activist groups offer support for families (like providing childcare at meetings), this kind of thing is by no means the norm.

Eli Pushkarewicz, who lives outside DC, told us that as a trans/nonbinary person, they had not ever felt the urge to biologically reproduce. They began to notice how the stories of "family" all focused on cisgender married parents, grandparents, and other biological ties. "Alternative families, including those whose parents were not married, single adults, queer families, or ones where the child's caregiver was not a biological parent, were always treated as if something wrong happened resulting in the 'not normal' family," Eli says. "There also was no space given to the idea that adults maintained their friendships or identities outside of work/family responsibilities, as if once they became adults, parenting became their sole reason to exist."

Eli and their partner ultimately became foster parents, providing a temporary, loving space to queer youth who are caught in the tumult and grief of the foster care system. But for other people in their life, this has meant a learning curve. Eli tells us that a friend recently confided that "he felt very unsure of how to engage with our other 'people.'" Eli takes time to choose their words: "As he has young kids, he was out of practice on how to engage groups of other adults in his age range. Whereas his life was centered around his children, the rest of us were focused on our relationships with each other [and] activities in the larger world. This reminded me of one of the beautiful aspects of the family structure we are building: it enables us to parent [our foster children] while continuing to nurture the other things that are important, as we are not just dependent on a small biological support network."

Climate-related reasons inform many people's decisions to have children, but even for those parents, the climate pressures and worries never go away. The decision to have children is not a discrete event but a process: the same pressures play out continually for climate parents as they do for nonparents, and in many ways that nonparents don't experience. Having children often raises the stakes of the fight—climate parents are caring for someone vulnerable whom they love, and whose future extends much further than their own into the age of climate consequences. Parenting, for many, is one of the best experiences of being human. And through their children, parents feel a new level of vulnerability.

Considering other parenting strategies: Why doesn't everyone "just adopt?"

In the face of the Impossible Question, a common bromide is that everyone should "just adopt." We've met many foster and adoptive parents through this work, and spoken with people considering these options. However, most concerns are the same for adoptive parents as for biological parents: when you love a child, you inevitably take on responsibility for their future.

For parents considering adoption, it's important to listen: not just to adoptive parents, but to adult adoptees as well. Adoption brings with it a host of justice-related considerations, including those related to identity. Most adoptive parents in the United States are white, and many are adopting children who are not white.[5] In 2011, nearly 8 in 10 adoptive parents of kindergartners were white, but 6 in 10 adopted kindergartners were kids of color.[6] In the past eighteen years, more than 270,000 children have been adopted from places such as China, Ethiopia, Ukraine, Russia, Guatemala, and South Korea, more than 60 percent of them girls.[7] Many transracial and transcultural adoptees report feelings of isolation, cultural alienation, and struggles with identity.

In November—National Adoption Month—of 2014, Rosita González, of the adoptee writers collective Lost Daughters, led a campaign called #FliptheScript to draw attention to the lack of adoptee voices in the public conversation. She critiqued the prevailing narrative that holds adoptive parents up as saviors, and wrote, "I feel marginalized as an adult transracial adoptee, until I am among other adult adoptees. We talk and listen. We are hungry for validation. We are our own village, and we want to help those youngsters who will grow up to be a part of this village. The important thing for a child is her sense of belonging."[8]

We bring this up not to speak *against* adoption, but to present some of the complexity that has arisen in speaking *about* it. There's a major difference between how people discuss adoption who are seriously considering it themselves, and how people casually tell others what they should do. We frequently hear (usually in Facebook comments sections) that, "There are so many kids just waiting for a loving home. Why doesn't everyone just adopt them instead?" *Instead* here means "instead of increasing the global population by one."

This approach needs some critical consideration. "Everyone should adopt" demands the question of who is "everyone," and who, exactly, "everyone" is adopting. Adoption can take years and be prohibitively expensive. Proposing adoption as a broad solution to environmental problems reveals class bias, in many cases racial bias, and an intrinsic acceptance of inequality. In this view, one group solves problems and the other creates them: the wealthy can adopt, and it is the poor who need adopting.

This brings up a broader question: what systemic factors cause children to need new homes and families? Policies that exacerbate poverty, here and abroad, are a major factor in children being separated from their families, either temporarily or permanently. Many infants enter the adoption system through US Crisis Pregnancy Centers and their overseas equivalents, organizations whose purpose is to offer misleading, incomplete information to pregnant people and to willfully deny access to abortion services. Historically, the United States has played a significant role in destabilizing other countries, which has led to a rise in "adoptable" children. Christian missionaries and organizations have also long imposed their own determinations about who is fit to parent and who is fit to adopt. Drought and famine are other factors that drive people from their homes and separate families: we can expect to see climate change amplify these effects. Only an ahistorical, apolitical view of this situation would offer adoption as a panacea without looking critically at our country's complicity in orphaning our own, and others', children.

Adoption is not inherently simpler, easier, or more correct than is biological reproduction. It is the right answer for some families, and can connect children and adults with the love and care that they need. But we can't mistake it for a universal answer to the Impossible Question.

CHILDFREE BY CHOICE

For many, climate change has influenced the decision *not* to have children. When reporting on climate and reproduction, the news media has focused the most on this group's existence, but not on the breadth of their reasoning. Both teen and adult birth rates have dropped in the past twenty-five years, and there's considerable evidence suggesting that climate concerns play a role in this decline.[9] But deciding not to become a parent does not mean you are a pessimist, nihilist, or that you don't plan to fight like hell. And no clickbait headline could summarize all the stories we've heard, unique declarations of resignation, conviction, anxiety, fear, and hope.

Some hesitate to give a child into a world that feels so unstable. On a summer evening in the stuffy basement of the Museum of Reclaimed Urban Space in New York, we recorded testimony from Hannah Harpole. The group discussion continued upstairs. Hannah, a thirty-four-year-old white doula with a serious gaze and a gentle manner, said: "I really love children, and I love babies in particular. . . . But I at this point feel that it's very unlikely that I will have children. . . . It's a feeling: like my body just observes when I go to the forest where my family lives . . . that things are changing and there's a problem, and it wouldn't be fair or a good idea to bring someone else into this change, this uncertainty."

Other people see their own decision not to have children as providing a bit more space for someone else's child to thrive, a sacrifice as hopeful as it is heartbreaking. Rachel Ries, from

Minneapolis, told us in a letter: "one day i realized the coloring of my world was a bit different and this difference came from knowing i likely wouldn't bring new life into it . . . i have no confidence i'd leave my daydream children or grandchildren with a hospitable home on this planet. and i can't abide that. and if my not bearing children tips the sustainability scales in favor of other children already here—even a teensy bit—then may it be so." Rachel had always imagined she'd be a mother, and her realization was painful but resolute. She viewed her decision as a sacrifice for a greater good: hope for the welfare of the children already near and dear to her, and a belief that she could not ensure a safe or healthy life for a child of her own.

Marla Marcum, an organizer in Tennessee, and her husband Jake have a related story. Marla is a seminary-trained Methodist minister who grew up the only child of a single mom in Missouri. She says that long before she came into a sense of her work in the world, she felt certain that she would have a daughter. She and Jake married when they were both twenty-five years old. Neither felt a rush to have a child. "I knew that I was going to work on climate change because I had a sense of the pressing reality of a coming crisis," Marla says. Over the next three years, she studied the moral and spiritual dimensions of movements for social justice. She faced—and moved through—her own despair about the people and creatures and places that were already lost, and the recognition that more destruction was coming.

By the time she turned thirty, Marla was questioning her dream of a daughter. "I was gaining clarity that the work to which I was called wasn't a job or a career," she says. "My work was a vocation, an embodied way of life that would pull me away from home and family regularly. I didn't speak of it to Jake, and I put that pain aside. I experimented with the work. Jake supported and celebrated me."

When Marla was thirty-three, they discovered that she had a rare infection that destroyed one of her kidneys. "I mention this," she continues, "because the loss of that kidney brought Jake to his own silent clarity that we would not be attempting a pregnancy. I sat him down the following year to tell him that I was certain that I could not be the kind of mother I would want to be, while doing the work that was mine to do in the world." Jake's emphatic response was, "Of course we're not having children! You have one kidney, and that's too dangerous. I won't risk losing you." Marla cried tears of joy, though she knew Jake would have been an incredible father, and he had always wanted the opportunity.

When Marla told her mother and grandmother about their decision, she based her story in the strength that each of them had given her. "I let them know that I was doing what each of them had very powerfully taught me about being a strong, loving woman—do what you must to make life better for the next generation . . . and I apologized that I would not be adding another daughter to the story. We cried," she says, "and they told me they were proud of me."

And Marla has become a mentor and a parent in other ways. She cofounded Climate Summer, an internship program that sent college-age kids across New England on bicycles to educate and organize for climate action. We were moved to hear from a program graduate—and full-time climate activist—Nick Jansen, how deeply he and his climate summer classmates care for Marla. She told the story of her decision not to parent to her Climate Summer mentees, and it landed hard on Nicholas, who was nineteen at the time. He blurted out, "But you're *our* mama!" "I became," Marla says laughing, "'Mama Marla.' I resisted at first because it stung a little, but I came to know that it was a true expression of love and connection, and so I embraced the title."

She grows serious. "I love children and teens and young adults," she says, "and I'm committed to being Mama Marla for as many of them as I can. There are moments when I am terrified of outliving Jake, being old and alone, and not having any children to care for and love me." In those moments of fear, she finds comfort in words she learned a long time ago: "Abandon attempts to achieve security, they are futile. Give up the search for wealth, it is demeaning. Quit the search for salvation, it is selfish. And come to comfortable rest in the certainty that those who participate in this life with an attitude of Thanksgiving will receive its full promise."

"I embrace this choice as a sacrifice," Marla says quietly. "'Sacrifice,' a word whose meaning is mostly lost in our time. To sacrifice is 'to make sacred.' It is a choice that nobody can make for me. It is to forgo something for oneself in order to live into a greater calling or purpose. And so while there is some loss or pain in the choice, it is a joyful embrace of life."

Reverend Mariama White-Hammond, a Boston-based climate organizer and the director of the city's sustainability office, told us that although she had already made the decision not to have children, it did not start with the climate crisis. "It started," she tells us, "with falling in love with a man that wasn't sure that he was ever going to get married, or have children." Although she had never thought of a world in which she *wouldn't* have children, "we came to this point where I had to decide: was this a dealbreaker for me?" As they got serious about marriage, Rev. White-Hammond had decided she was clear enough that "this is the person who is my life partner, that I had finally come to the place where having children is not something I have to do." She takes a sip of her drink. It's midafternoon and we are talking at a hotel cafe near UMass Boston's campus. She continues,

> But part of that is deeply couched in the reality that I grew up in a system in which family was very strong. I've always had my biological family, but I've always had a family significantly larger than that. For instance, my dad had one sibling, my mom had seven. My mom's family moved all around because my grandfather was in the Air Force. There weren't that many Black people in the Air Force, so they'd end up in a lot of places where there's not that many Black people. And because they had moved everywhere, they kind of manically made Boston

the family hub. So five of the eight kids settled here. And I grew up with a lot of my cousins and one of my uncles went on to have seven children, who all started having children relatively early in the game, and so we had this humongous family.

Rev. White-Hammond describes an expansive family that offers lots of people to love, and lots of people who love her. We will return to this vision of family in chapter 5. Though their stories are each unique, Rachel, Marla, Rev. White-Hammond, and others like them all commit some of their energies to the children who already exist, and to those who will.

But *not* having children, like having them, is a process rather than a discrete event. It unfolds over the better part of a lifetime. For instance, people without children interrupt a number of cultural narratives and patterns, and are often seen as outside the norm. This creates pressure that women feel especially: time and again, women at our gatherings worry that if they don't become mothers they will be seen as, or feel like, "less of a woman." One older woman remarked that reaching a certain age was a relief because people had finally stopped telling her she might change her mind. These pressures are most common among heterosexual women, but not rare among queer women either.

It is, however, a gendered pattern. Men who don't parent are still generally considered "full" men—not family men perhaps, but they don't lose too much of their social legitimacy in not becoming parents. Motherhood is still entwined with powerful social expectations for women that we often feel deeply, and respond to. Motherhood-as-total-fulfillment is baked into our cultural narratives a thousand ways, from Marmee to Lorelai Gilmore to social media as the arena for maternity and baby picture competitions. Women without children are often depicted as flawed or cold; at best kind of lonely and pathetic, at worst, unnatural.

And while we've often spoken with people who are struggling to decide, the decision not to have children is not always

PART I

difficult. Some people just don't want to. Climate concerns may or may not be a factor here, but we want to point out explicitly that choosing a childfree life is not inherently nihilistic. For people who strongly desire children, this can be hard to understand. But being child-free can be its own reward, enabling people to devote their lives to other things—including, importantly, social justice, and creative or professional work. A drawback that many in this position report is that it's hard to find acceptance. Some child-free people have reported incredulous or dismissive responses from friends and family if they cite climate change as a factor, and so they may keep it to themselves.

Being childfree in a culture that sees parenthood as a default setting can leave people at a loss for how to pass on their gifts to the next generation. This is true both for people who wanted children and didn't have them, and people who didn't want them to begin with. Where do we put our love, our hope for the future, if we do not have children "of our own?" Mainstream culture in the United States, with its limited view of community, huge emphasis on the nuclear family, and heightened alert to "stranger danger," offers few new solutions to this problem. We have few opportunities to interact with children unless we happen to be in a kid-focused profession like teaching, or to have family close by.

So much of US culture—from nursery rhymes and old stories to donut recipes and kitchen tools—is handed down within this narrow conception of family. Not having children can leave people feeling that their culture has reached the end of the line, as if it will die with them. This rift between generations can increase a sense of alienation and loneliness. Many of our participants, especially those from middle-class backgrounds where a small and geographically dispersed family is the norm, identify this as a piece of their struggle. We believe the first step toward healing this rift is a more inclusive and robust understanding of family, which we delve into in chapter 5.

But having forgone the map of tradition, childfree young people are blazing unfamiliar trails, and cultivating kin relationships with aspects of the natural world. Twenty-one-year-old Claire grew up in a family of falconers. She describes it to us at a house party in Amherst, Massachusetts: "That's a practice where you take a bird in for a season, and you release it. That's a really hard relationship. But by taking a bird in for a season, you are increasing its chances of survival by 60%. And it's really frustrating and hard at times, raising a bird." She shifts on a low ottoman. Outside is the gentle chatter of voices, of people serving themselves tea. "This has made me wonder," she says, "whether you could have the same kind of gratification by having a relationship with a hawk, that you can with a kid." Claire takes a breath. "And I hope that you can. Because I'm not planning on having kids, and I think that I need something in my life to fill that space."

There is no single profile for the growing number of people who decide not to reproduce or parent. For some, it is a straightforward choice. For others, it's a heartbreaking but necessary decision based on a personal reckoning with the future. Some have made a considered decision to relinquish the space and resources their child would have occupied to other children, who *will* be born. Some are teachers, caregivers, piblings, or other kin whose needs to be around the young are adequately met in other ways. And some have decided that the time and passion required to fight the climate crisis leaves too little to parent a child. There is no one story that shows the whole range of climate consequences in our lives.

THE UNDECIDED
The majority of people who come to Conceivable Future events are still in the throes of these decisions. How we weigh our possible futures can reveal hidden aspects of the climate crisis, shed light on the ethical dimensions of history, and point to possibilities for the future—but only *when we ask the right questions*.

We consider the conflict that is apt to arise in a hotter, more unstable world. "I'm afraid that [my babies]'ll have to live in a world that is increasingly full of war. I'm afraid that they'll eventually have to live in a world where there is no fresh water," says Meghan Hoskins in a testimony recorded in New Hampshire. Meghan, a redheaded white woman with stylish glasses, came to a conversation we held in a renovated historic mill, along with a dozen or so other young activists and a handful of women from a local mothers' group. "But I'm also deeply afraid of not giving my parents the beautiful, little, red-headed grandbabies that they've always wanted." She is speaking through tears.

Meghan's fears are grounded in geophysical fact: water is drying up in many towns and cities. There is abundant evidence that an increase in global temperatures is leading to an increase in conflict around the world. These include large conflicts such as wars that erupt out of famines, for instance, but also smaller interpersonal conflicts.[10] For each degree Celsius of temperature increase (1.8 degrees Fahrenheit), Stanford researcher Marshall Burke estimates that there could be a 20 percent increase in civil conflict around the world ("civil conflict" includes crimes like assault and robbery, but also road rage and fights at sports games).[11] Climate change is also linked to increasing suicide rates.[12] It seems that all of human life is calibrated to our current temperatures—including our emotions.

Other young people are focused entirely on what needs to be done to prevent all-out catastrophe, shunting reproductive decisions for now. "One of the reasons I got involved in the climate movement is I feel like there's a really, really small window that we're working within," says McKenzie, also from New Hampshire. "So it's our calling to be part of it. The same way I see what happens beyond this window as an abstract thing that I don't deal with because I'm in the here and now, I think of . . . that decision to have kids as beyond that window."

In our discussions, participants are frequently conscious of the ways that the climate links to other struggles around reproduction. When we spoke to Leah Quimby, the Zika virus was making daily headlines by causing irregularities in pregnancies and births in Central America and on the US border. Leah was concerned about how Zika might couple with increasing restrictions on reproductive healthcare. "Now we have Zika virus coming into Texas," she said, "and in the meanwhile you have the government cutting down on abortion, birth control, and access to reproductive healthcare in Texas, leaving people in a real bind." When fetal health is imperiled by a climate-intensified disease, and access to reproductive healthcare is imperiled by a restrictive government at the same time, the dilemma for pregnant people is Kafkaesque.

For Amelia Keane, age twenty-three, this is a question of human rights: "I have a choice to have kids or not, and I know that a lot of women don't," she says. "I think it's selfish at this point of me to want to continue bringing children into this world. But it's a conflict—because that should be everyone's right, and I feel like in a way it's being taken from me."

Amelia's comments reveal the distortion at the heart of this experience for so many young people. And despite that self-judgment, she recognizes that having children *is* a human right—a right to which she is entitled, that is unequally accessible and threatened by decades of bad leadership and unchecked corporate self-interest.

Young people often get pressure from family, friends, and even strangers around reproductive decisions. The tone varies from loving to nagging, or even accusatory. It can mask the depth of grief and fear that many of our Elders feel, knowing that their children face a far less certain existence than they themselves ever did. Some older adults may find it easier to exert pressure on adult children to fall into generational line, rather than to listen to their concerns. In Washington, DC, climate organizer Camila Thorndike describes one such bad interaction:

An older man told me "well, throughout the nuclear era we were all hiding under our desks and women were wondering whether it was the right thing to have kids then." And I was so pissed at him! Because this is completely different! That's one person's choice, whether to press *one* red button. But what we deal with now is so many small red buttons . . . and more than that, in the lifetime that I've had, with decision-makers and scientists and my own parents knowing that this was coming down the line. And *still* we are making the kinds of collective decisions we are making. That's different!

Camila isn't invalidating past or present concerns about the nuclear threat, but asking her Elders to recognize that her climate concerns *are also valid*. As she points out, nuclear proliferation is a political problem that dirties far fewer hands than the climate crisis. The harm from greenhouse gases, however, is already underway, and will inevitably increase from the runaway geophysical processes already unfolding, even if we stop using fossil fuels today. The most committed political action can only alter the upper limit of harm, not the baseline.

Camila's testimony also points to intergenerational tensions we're all familiar with. For many of our Elders, the world was predictable in certain ways. Baby boomers with a bachelor's degree held, on average, eleven jobs over the course of their lifetimes, and were employed, on average, 78 percent of the weeks from age eighteen to age fifty.[13] They paid into Social Security, and upon retirement they have been receiving checks. In the background, from the 1940s till the 1980s, the seasons played on repeat. By contrast, our generation usually patches together a living wage without benefits, from two or three gigs at a time. We are expected to hold approximately twenty jobs over the course of our working lives, if we're lucky enough to be employed.[14] We also pay into Social Security, although we don't expect the money to hold out for us. And if we were born in 1985 or later, we have never experienced a colder than average month on earth.[15]

It can be shocking for our Elders to learn that even their basic assumptions about life simply won't hold up for us. "I can't even begin to think of a 'normal' life with this hanging over us," says Marvin Warren, a white nonbinary permaculturist and botanist who lives in the Hudson Valley. For some older adults, it's difficult to fathom the ways in which this crisis is so present for their own children. It can be painful, and even humiliating, to acknowledge that decades of corporate chiefs and neoliberal policymakers betrayed their children as a matter of course while they were promising stability. This may be especially hard for members of the historical counterculture, hippies among them: it requires acknowledging major defeats.

When our Elders dismiss our worries it may come from a desire to protect us, but the impulse to pretend that things will be okay is hurtful. As Camila mentioned, there were people who, during the nuclear crisis, had been asking similar questions to our own: is it safe to have a baby? Will my child grow up in a hospitable world? The fear was pervasive enough that Bob Dylan sang about it in "Masters of War." In the early days of Conceivable Future we were excited to speak with them, they seemed like natural allies in our organizing—people from whom we could learn.

But some of our interactions with these folks were profoundly different than what we had envisioned—and profoundly disappointing. For many women of our mother's ages who also had concerns, the fact that the world *didn't* erupt into nuclear warfare was used to dismiss our worry about the threats that we face, now. "We went ahead and had children and it was the best thing we ever did," they told us. "Things'll probably be okay for you. Don't worry so much."

Nobody likes having their concerns dismissed. We didn't—and history certainly confirms that powerful groups are prone to dismissing the objections of less-powerful ones. These exchanges, though, were unsettling for reasons bigger than our hurt feelings: they represented an unwillingness of an older generation to empathize with a younger one. The fact that the world had ignored *their* concerns seemed to empower them to dismiss ours, instead of providing a foundation for solidarity. Any successful intergenerational movement needs to take the realities of each age seriously, and to respond to them with unity. As journalist Gayle Golden writes: "young people wrestling with these decisions need exactly what the best Nonnas, Oba-chans, Abuelas, Babus and Nanas have done through the generations: quietly shown up without fanfare and created opportunities for conversation by simply listening to those fears and hopes."[16]

If we could summarize a decade of these conversations in one observation, it is that we conceive of our lives and futures through our loving relationships. Life is worth living because we love each other; concerns about bearing and raising children are also based in love, and they reflect some of people's deepest fears about the future. Thirty-one-year-old divinity student Kathy Lin put it bluntly and powerfully: "Do we think that life is worth living or not? And if we think that life is worth living, having a child is an act of faith. As my decision to keep on living, to step into tomorrow, is an act of faith. Do I think life is worth living or not? And if I don't think so, it's a really sad position. I experience in my emotional state this question of 'is life worth living?' And my answer to that is, frequently, 'I'm not sure.'"

Kathy is naming the connection between what we live for and what we have to fight for. The negative formulation—that losing faith in the future makes it hard to live—also has a positive counterpart: what gives us joy—the good in life—can be in *how* we fight, as well as what we fight for. No one makes reproductive decisions freely when the future is being used as a sacrifice zone. We are learning to live with imperfect answers and a terrifying level of uncertainty, but if we want a humane future, it's on us to ask the bigger questions and challenge the larger institutions that make the world so inhospitable to *everyone's* children.

This conversation is bigger, more inclusive, and has far more nuance than it did when we began our work. But it's still just beginning. We will keep telling these stories, and more, as tools for change. Our stories have the power to:

- Change the narrative from individual to collective focus
- Challenge the assumption that we must be experts to care or speak about the crisis
- Create common cause with others
- Illustrate what we don't want
- Conceive the future we *do* want

It is a human right to maintain personal bodily autonomy, to have children, and to not have children. If people don't want to have children—for any reason—that is also their human right. But if their reason for not having children is because they've internalized the burden of guilt for the carbon footprint of a potential child, then we have a different recommendation: put that guilt on the doorstep of the fossil-fuel profiteers, and of the politicians that enable them.

CHAPTER 1

A Dangerous World for Birthing

THE CLIMATE EMERGENCY IS STILL WIDELY DESCRIBED IN future terms, and illustrated through satellite images of shrinking ice caps. But we don't need to look to the future, or to the poles, to see it; it's *already* harming people's bodies, families, and communities. It is already amplifying inequality. We know how bad it is getting. And this isn't just an external force bearing *down* on us. Rather, it's moving *through* us, revealing how interconnected we are with the world around us, at the cellular as well as societal level. Heat and pollution's effects on pregnancy and child development show us how porous we are. We *are* nature. For people planning families, the climate crisis is a burden of negative factors, measurable already, intensifying with time to an unknowable extent. What is the present landscape of our reproductive health, and what will we and our beloveds experience as the temperature rises?

We spoke with two experts in this growing field to help us unknot this tangle of harms: Dr. Adelle Monteblanco, professor of public health at Pacific University, whose focus is the effects of extreme heat and disasters on pregnant people and youth, and Dr. Sandie Ha, an environmental and perinatal epidemiologist at University of California, Merced. We rely on their insights throughout this chapter to illuminate the big picture, but we begin with a testimony from thirty-four-year-old organizer, artist, and translator Yudith Nieto. Yudith became a youth organizer around 2012, during the Keystone XL Pipeline fight. That struggle connected activists "from Canada all the way down to what we call Cancer alley, and the Gulf South," she explained, "where we have grown accustomed to co-existing with all of the industries, refineries, tank farms and pipelines in our neighborhoods."

We interviewed Yudith about her organizing work—initially planning to introduce her later in the book—but the story of how she began to organize is also the story of her personal experience of the health hazards of fossil fuels. Born into a farming family from San Luis Potosi, Mexico, Yudith and her family left their traditional land for the United States after NAFTA through the Bracero program. They ended up in the East End, one of the most polluted communities in the city of Houston, Texas. "I was thrown into environmental justice work because of our need to fight for the right to clean air and clean water and clean soil," she tells us.

> I came from very natural ecosystems—very beautiful landscapes, being able to play in the water—to one of the most dangerous communities, where you smell all kinds of chemicals in the air. We have at least 10 different known carcinogens in the air of the community of Manchester, which is in the East End, where I grew up, day and night.

Yudith came of age in that environment, watching her little cousins suffer from asthma, her aunts experience "issues in their reproductive organs, reproductive issues, their children being born early, and having children with mental health and development issues." In addition to spurring her toward environmental justice, these experiences informed her own thinking about whether or not to have children, about sovereignty over her own body, and about her commitment to reproductive rights.

Because all of these chemicals, all these industries, were the reason why my lineage, my people, were now experiencing newly found reproductive issues which we hadn't seen before in our families. And then they've had children in those environments, developed as young women in those environments. Now, we're having issues having children, and that let me know that I too may have a child with special needs, or who may not even survive. I may have miscarriages. And so when abortion rights were abolished in Texas that really spoke to me, because not only do I not have the right to say "No, you cannot have a permit to emit even more toxic chemicals into the air," but I also now can't say "I cannot carry this child to term knowing that they will have development issues because of the industries that I was exposed to." So, yeah—that's a lot there.

That is a lot, and it moves across generations, concentrating over time. Dr. Monteblanco agrees succinctly, "The climate crisis shapes every stage of our reproductive lives."

THE BIGGER PICTURE
In 2022, Dr. Sandie Ha made a major review of the available scholarship on how global warming affects pregnancy and infant health outcomes. Her research analyzed more than seventy recent studies, looking for patterns. She told us: "Although the impacts of climate change on health have been raised previously, it is only in the most recent IPCC report that pregnancy health was linked to climate change." For a little context, the Intergovernmental Panel on Climate Change (IPCC) is a United Nations–led group of scientists that has been tracking the most up-to-date research and data about the developing climate crisis since 1988. They create comprehensive reports with science-based recommendations to world governments, which have gotten starker and more alarming with each publication.

Dr. Ha continues, "When discussing health impacts of air pollution, for example, most people think about people with respiratory diseases, children, or the elderly. Thus, we need to raise more awareness regarding the vulnerability of pregnant people in the changing climate."

She organized her findings into three categories of climate-fueled harms:

- Direct Impacts: things like heat waves, wildfires, or extreme weather events.
- Indirect Impacts through the physical environment: air pollution, food, or water scarcity, changing vectors, hosts, and pathogen distribution (e.g., contributing to things like pandemics).
- Indirect Impacts through the social environment: food and water insecurity, conflicts due to resource scarcity, and displacement.

These three kinds of harm combine in people's environment to create biological and physiological changes (things like inflammatory responses, endocrine disruption, malnutrition, and infection), as well as psychological changes (like stress and mental health problems).

This whole system of worsening conditions combines with preexisting social and environmental problems, as well as our dysfunctional healthcare system, to create adverse pregnancy outcomes:

- Gestational complications
- Low birth weight
- Restricted fetal growth
- Preterm birth
- Spontaneous abortion
- Neonatal mortality

These findings, which we'll explore in depth throughout the chapter, reveal a grim picture of the hazards we are *already* encountering, and of the ways that rising temperatures, contaminants, disasters, and inequality are *already* expressed in human bodies. But here it's worth pointing out, before we get into the health data weeds, that these health problems arrive on a landscape that is already unjust to parents, pregnant people, and children. The United States is already an anti-natalist country in everything but name. Our policies have made reproduction unnecessarily dangerous, difficult, and expensive, and have devalued the lives of children in this country. Now the health effects of climate change render these dangers all the more severe.

Antinatalism in the United States

Rising temperatures are causing and exacerbating the harms that flow through our bodies, our homes, neighborhoods, the air, and water. But in order to fully understand the reproductive terrain Americans are confronted with, we need to name some of the factors that *already* place the United States among the worst countries for parents and children in the global north. These factors include lack of access to prenatal and general healthcare, rampant gun violence, lack of parental leave, job insecurity, and racism—and the latter is especially visible in disparities in maternal and infant health.[1]

"For decades now, the U.S. has had dangerous rates of preterm and low-birth weight infants," Dr. Monteblanco tells us. "Importantly, the risk is higher for Black birthing people and Black infants, and these inequities were/are compounded by [the COVID-19] pandemic." March of Dimes assesses the country's preterm birth grade as an appalling D+, making the United States an outlier among wealthy countries.[2]

Low birth weight and preterm delivery are just the beginning. After children are born, they continue to live in a regime not designed for their well-being: daycare costs more than college in many places, and "childcare deserts" continue to drive women out of the workforce.[3, 4] The minimum wage is not a living wage, and healthcare is still the top cause of bankruptcy in the United States.[5] Teenage girls across the United States are "engulfed in a growing wave of violence and trauma," according to federal researchers, including sexual violence, anxiety and depression.[6] We are experiencing a national—and racialized—sleep and rest deficit, which produces a host of negative consequences for people forced to drive their bodies unmercifully in order to survive.[7] Psychology professionals tell us we are in the midst of a mental health crisis, intensified by the isolation and trauma of the pandemic. We understand it to be downstream of the crises we're naming here: emotional suffering is the inevitable consequence of relentless hardship with inadequate access to care.

These and a range of other discriminations *already* endanger health and safety in the United States. This is *already* an unfriendly place to have a child, especially considering what social goods the United States could conceivably support with all the wealth at our disposal. When young people consider whether or not to have a child, this inequality is *already on the table*—the climate crisis arrives on a landscape that is already unjust and hostile to families. No one is guaranteed a safe, healthy, or long life for themselves or their offspring. Human existence has always been risky. But while this country has been leaning into inequitable risks on a policy level, the additional harms caused by temperature rise are broad, unmistakable, and new.

For Dr. Monteblanco, reproductive justice is the framework that best allows her to express the stakes of her work in this field. "The climate crisis threatens reproductive health outcomes and a commitment to reproductive justice," she tells us. "Reproductive health outcomes include preterm and low birth weight. Reproductive justice, as defined by [the organization] SisterSong is 'the human right to maintain personal bodily autonomy, have children, not have children, and parent the children we have in safe and sustainable communities.'" Dr. Monteblanco likes to make a minor addition to such a thoughtful and visionary definition, by including the ability to "*gestate* and parent the children we have in safe and sustainable communities." In chapter 2 we'll talk more about reproductive justice as an intellectual framework and as a movement. For now, we're defining some of the problems that it seeks to solve.

Many of these problems are, at their core, political and social. Political instability tends to negatively impact reproductive health and sovereignty, even when reproductive rights are not the explicit target. In other words, the Trump years were already bad for reproductive health, even before the *Dobbs* decision was handed down. 2022 was the year Americans were stripped of their legal right to abortion access nationwide. This was bad news across the country for people who need abortions, and it's doing catastrophic harm to people in the states that have banned abortion, especially minors and those without the social or financial resources to travel. Among those in need of abortion access, trans men and nonbinary people are most often excluded from appropriate reproductive healthcare systems already, and in an era of increasing repression toward nonconforming bodies, and lives, these fragile healthcare systems are more imperiled still.[8]

This is the political climate we're living in. These risks reach each of us differently: they are not evenly distributed, but that big picture shapes health outcomes for all of us.

LIVING WITH DIRTY FUELS

We know this is a bad time for reproductive health in the United States, and we know that pollution and extreme heat and weather events hurt us, but there's a missing link here in the public discourse.

Wherever fossil-fuel extraction, processing, or combustion takes place, people's health is hit the hardest. Sacrifice zones surround every coal plant, chemical factory, fracking well, and oil refinery, and include Yudith's neighborhood, Manchester, in Houston's East End; Port Arthur, Texas; and "Cancer Alley," the grim nickname given to a hundred-mile stretch of the Mississippi river on which the density of oil infrastructure has created an epidemic of disease for residents. The advent of extreme fossil-fuel extraction, such as fracking and tar sands mining, have exposed more regions and more groups of people to the dangers of fossil fuels. Fracking wells dapple the country, and pipelines crisscross it. The edges of sacrifice zones are blurring.

Living near Liquefied Natural Gas (LNG)

Mercifully, few of us believe the PR that natural gas is the "clean-burning fuel," a bridge fuel, a lower-emissions fuel, or any related lie. We've probably all seen the flaming tap water, and heard about the little earthquakes hydraulic fracturing (fracking) causes, but we may not be familiar with LNG's full rapsheet. Many of the hazards we rightly associate with fracking are what happens when fracking goes wrong: pipeline explosions and chemical leaks that poison aquifers. Satirical paper of record the *Onion* immortalized these grievances in 2015's "Pros and Cons of Fracking." Pros included:

- Blasts tens of thousands of gallons of chemicals deep underground, out of harm's way.
- Prompts important conversation about whether or not people have a right to clean water.
- Fact that shale well blowout could happen at any moment emphasizes ephemeral beauty of life.[9]

But when a fracking well is working as it's intended, it's already increasing pregnancy complications, causing premature births, neurodevelopmental birth defects, and respiratory problems in infants and children in the surrounding area. By plotting over 10,000 pregnancies around active fracking operations in Pennsylvania, researchers from Johns Hopkins Bloomberg School of Public Health "found that living in the most active quartile of drilling and production activity was associated with a 40 percent increase in the likelihood of a woman giving birth before 37 weeks of gestation (considered preterm) and a 30 percent increase in the chance that an obstetrician had labeled their pregnancy 'high-risk,' a designation that can include factors such as elevated blood pressure or excessive weight gain during pregnancy."[10]

A separate study, published in *Reviews on Environmental Health*, identified five *categories* of "pollutants of concern," which cause problems they've divided into three large groups:

- Neurodevelopmental—the structural differences visible in the developing brains and bodies of exposed kids, such as neuroinflammation and small head circumference.

- Neurocognitive—the cognitive outcomes of that developmental interference, like deficits in memory, attention, and IQ.

- Neuropsychological—the emotional and behavioral results of those changes to structure and cognition, as in "an increase in symptoms of anxiety, depression and inattention," and "increased risk of ADHD diagnosis and other childhood psychopathologies like OCD and Tourette's syndrome from impaired self-regulation."[11]

And while gas wells spring up wherever bedrock contains frackable gas, LNG companies follow Ye Olde Environmental Racism playbook when zoning infrastructure to refine, compress, and transport it. When the New England energy company PSEG moved to replace its coal-fired power plant with a gas compressor station in Bridgeport Harbor, Connecticut, local activists protested the move. "Environmental racism is when an unfair share of pollution is placed on communities of color and low-income neighborhoods. That's what is happening in Bridgeport. PSEG is making it worse by trying to open a new gas plant, which would continue to release pollution in the air for decades," said Gabriela Rodriguez, a nineteen-year-old Bridgeport resident and a member of Capitalism vs. the Climate.[12]

There is no such thing as a "clean" fossil fuel, and every time infrastructure is developed to extract, refine, or store such fuels, business and political leaders sacrifice a community to make it happen. They often do so while gaslighting community members themselves over their own environmental behaviors and individual carbon footprints.[13] Justice requires that we shutter this infrastructure and end the extraction that, by its very nature, injures both the land and the people who inhabit it.

Air pollution and risk of miscarriage and premature birth

In her analysis, Dr. Ha found air pollution to be one of the biggest negative forces on reproductive health: "Studies have shown that air pollution and extreme temperatures, both of which are consequences of the changing climate, are associated with decreased fecundability, or the ability to become pregnant after a couple start trying. This impact may ultimately contribute to decreased fertility, which has been observed in both human and animal studies." Some of the biggest harm comes from the smallest measurable particulates in our air, the $PM_{2.5}$, particles, whose diameter is smaller than 2.5 micrometers.

Exposure to air pollution increases the risk of preterm birth, especially for pregnant people with asthma—a condition already linked to air pollution, and that already increases the risk of preterm birth.[14] This is a vicious feedback loop. Air pollution has also been shown to correlate generally with higher rates of miscarriage, and researchers in and around Salt Lake City, Utah, also linked short spikes of certain pollutants to spontaneous pregnancy loss. They found that "a 10-[part-per-billion] increase in 7-day average levels of nitrogen dioxide was associated with a 16% increase in the odds of spontaneous pregnancy loss."[15] And although the unique bowl-shaped topography of the city causes smog to sit on it like a lid, its average air quality is similar to most urban environments.

China recently bore up with equally grim evidence; spikes in fine-particle air pollution were found to increase under-five child mortality by more than 1 percent in a study that examined the deaths of more than 60,000 young children nationwide. A 1 percent uptick in child mortality is a cold little signifier for an epidemic of families grieving. "Mortality from diarrhea, pneumonia, digestive diseases, and preterm birth were significantly associated with exposure to $PM_{2.5}$," the researchers concluded. The effect estimates were larger for neonatal mortality, female children, and in warm seasons.[16]

Most of the causes of these small-particle spikes are directly or indirectly fossil-fueled: smokestacks and tailpipes, wildfires or controlled burns, dust from droughts—as in dry lake beds—lifted by winds. And while the closer we live to fossil-fuel plants the higher these daily small particle concentrations are, we can also see harmful concentrations of $PM_{2.5}$ in our homes with the small-scale combustion of our gas stoves and wood fires.[17, 18] Air pollution and climate change are interlinked harms, manifesting in our bodies.

Climate Harms
Heat

Even far away from local sources of pollution, fossil fuels are measurably harming people at every stage of their reproductive lives: they are doing so through heat. Two studies of tens of thousands of summer births in California show that when the ambient temperature rises by 10 degrees Fahrenheit, risk of premature birth and stillbirth rise by 8.6 percent and 10.4 percent, respectively.[19,20] Recent research has also suggested that heat contributes to the risk of miscarriage.[21] When we consider just how pervasive and harmful heat events are becoming, the numbers are terrifying indeed.

Dr. Monteblanco, who has studied many varieties of disaster, tells us that heat is a unique culprit. "Extreme heat stands out among climate-driven disasters because it is the number one weather killer, according to the National Weather Service," she says. "Every disaster reveals and exacerbates inequality. Folks marginalized prior to the disaster are hit with harder impacts. This is evident in heat preparation and response. It shapes [whether] you can afford air conditioning, if you can take time off work, how hot your neighborhood is, if you have to wait for the bus, how close you live to the hospital, and if you are insured."

Her words are evidenced all over the country. During the Chicago heat wave of 1995 over seven hundred people died, most of them low-income Black seniors who lived alone. In an NPR interview about his book on the tragedy, *Heat Wave*, Eric Klinenberg discussed the city's lack of emergency response, and came to this conclusion:

> What I came away understanding is that this was not a natural disaster. This was a disaster of our own making that had everything to do with race and inequality and isolation and things that still plague Chicago today. And if you think for a second about what happens when a hurricane is coming towards the city, you know, we stop the regular news programming. We have color-coded maps. Everyone braces for the storm. Political officials, if they're away, come back to manage the crisis. We don't do that with heat. And so, as the heat was approaching Chicago, meteorologists were frying eggs on the sidewalk.[22]

Dr. Monteblanco says that even twenty-seven years later, the inadequacy Klinenberg observed is still characteristic of official responses to heat and that overall, cities and even states behave as though heat is unpredictable and surprising, when it should be neither. She tells us that heat differs strongly from hurricanes, tornadoes, and floods because heat-related illnesses and death are so much more preventable than outcomes from other disasters (e.g., earthquakes). She says, "We *know* when extreme heat is coming, sometimes days or weeks in advance. We know who is vulnerable and we know how to keep people safe. We can mitigate and adapt, it's absolutely in our reach. But unlike other disasters that create a lot of property destruction, people do not take heat as seriously as a threat—for their health or their community." Heat is slow-moving but serious, she says. "Humans think their bodies can adapt faster than is often the case. But especially if you are working outside, or you are taking certain medications, your risk is higher and heat needs to be taken seriously."

And she reiterates that inequality plays a big role in the consequences of extreme heat: "To be clear, it's not just a climate-driven disaster. Like other disasters, it's shaped by decades or even centuries of local and national policies that contribute to the hazard and human vulnerability. Neighborhood exposure to heat is shaped by zoning laws and redlining, not just climate change." She points as well at zoning laws in some states, which do not protect renters' rights to use window AC units. "I think about the heat dome that the Pacific Northwest experienced in 2021," she concludes. "Certainly anthropogenic climate change intensified the heat wave, but the disaster is so much more than that."

Part of preparedness are the stories we tell about what is happening and why. In Dr. Monteblanco's view, the media isn't communicating the real dangers of heat, or identifying the state and federal policies that contribute to heat risk. "The media has been heavily criticized," she says, "for its use of images in discussing heat events (e.g., photos of kids playing in sprinklers or melted ice cream or even a pregnant person enjoying a day at the beach), rather than images that communicate the dangers of heat exposure."

Heat events will not stop—they are becoming more frequent. Because of the climate crisis, a child born in the United States in 2022 will experience thirty-five times more life-threatening extreme heat events than one born about sixty years ago.[23] Heat is a disaster, but also an accelerator of other kinds of disaster. And like the climate crisis itself, disasters are largely unnatural: a consequence of the social and political priorities within a society. "Disasters are not freak accidents," writes Samantha Montano in her book *Disasterology*, "they are the inevitable product of the decisions we, or some people, make."[24]

Preparedness will by necessity become our way of life. In early 2020, many people prefaced their COVID-19 preparations by saying something like "This might be crazy but . . ." It is not crazy to plan for a disaster. It's crazy to wait *until* the disaster to plan. And while that "we" has long been our fumbling leadership and powerful corporate interests, "we" can also refer to us as people, who—when organized—can demand better.

Disaster and vulnerability

As the temperature climbs, we are experiencing other types of disasters with greater frequency and intensity—these range from floods and superstorms to windstorms, wildfires, and pollution spikes. And these events, in addition to being terrifying in the short term, test communities over the medium and long terms as well: climate-driven disruptions to ordinary life can cut people off from any support they depend on.

An average of 21.5 million people have been forcibly displaced by weather-related, sudden-onset disasters and hazards—such as floods, storms, wildfires, and extreme temperatures—every year since 2008.[25] Disasters are riskiest for groups that are already vulnerable: infants and children, the pregnant, the elderly, people with disabilities and chronic illness, people escaping abuse. These people depend on caregivers and support systems that are often already overstretched and underfunded, like shelters and health clinics. In emergencies, such networks are often conscripted into service for the general population. At the same time, states of emergency resulting from storms, floods, and droughts create compounding threats like political instability, that frequently curtail human and civil rights.

Disaster-driven healthcare disruptions run the gamut, and include basic reproductive healthcare and family planning services. Disasters are also gendered—they affect women more adversely, including by more reproductive tract infections, early onset of labor and infertility, increased number of children, and shortened life expectancy.[26,27,28] Dr. Monteblanco points out that even something as basic as menstrual health is threatened by disaster: people who menstruate may have limited access to products or report infection (due to lack of water and supplies) during or after an acute disaster like a hurricane, but also in chronic disasters like drought. Disasters also shape access to birth control, especially for people who evacuate away from their homes and communities, as was evident after Hurricane Ike.[29] There is also increased danger from sexual assailants after a disaster, often in evacuation centers or temporary shelters.[30]

No matter where and when a disaster hits, there are likely to be pregnant people affected by it. Disasters, because of infrastructure limitations or hospital demands, frequently lead to delayed prenatal care as well as trauma. Lack of access to a doctor can interrupt a parent and child's nursing. And there is increasing evidence that experiencing a disaster raises the risk of postpartum mental health difficulties.[31] It's grim news for all of us.

The spread of disease is also more likely in the context of disaster, as unsanitary conditions can arise following hurricanes and floods, and even non-water-based disasters. Whenever hygienic infrastructure is damaged and must be improvised, and particularly where there is standing water, the risk of waterborne parasites, skin and soft tissue infections and diseases rises, as well as the risk of mosquito and other water-breeding, insect-borne illness.[32]

All of these diseases pose specific risks to childbearing, but Zika is of specific concern. The Zika virus, first recorded in humans in 1952, is primarily spread by infected *Aedes* mosquitos. While an infected adult may not experience severe symptoms from Zika, the virus is known to cause cases of severe birth defects in babies—particularly microcephaly, or a very small head—when their pregnant parent is infected. The range of Zika used to be limited to places with tropical climates. Now, many more places meet that description. In 2016, the CDC issued a no-go warning for a one-mile radius area in the city of Miami because of the presence of Zika-infected mosquitos.[33] This is increasingly common across the Gulf Coast and the US South.

"The climate crisis is a global one, but its impacts vary by place," says Dr. Ha. She describes her own experience of the contrast between where the harm originates and where it is felt. In Vietnam, she and her family lived in a small village in the central part of the country, which barely generated emissions. However, the region endures extreme weather events every year, and the events are worsening over time. "We would have to rebuild our thatched house annually after a hurricane passes, leaving not only broken infrastructure but also often infectious diseases," she tells us. "When I moved to the San Joaquin Valley, the issues we're facing include wildfires, droughts, and extreme temperatures, all of which seem to be objectively worse every year. I also see the same trend where groups who are already burdened by other issues (i.e., poorer, less educated, immigrants, etc.) are more impacted." At the same time, she says, communities who are most impacted and know about the solutions that work best for them are often left out of the problem-solving conversations.

Male health and the global decline in sperm count

Dr. Monteblanco points to climate/reproduction as an anomaly in medical studies: in this area alone, there is more research about female bodies than male. This tracks tidily with the American national obsession with policing female bodies, but the point remains that the effects of climate and heat on male health are still understudied. "I'm really curious about male bodies, reproduction, and climate," she tells us. "Women are subject to endless advice and guilt about healthy pregnancies, but men's bodies, including sperm production, continue to be overlooked. We know little about how their bodies matter for reproduction—and how their bodies might be shaped by the climate crisis—including but not limited to pollution, heat, allergies, etc."

What we *do* know, though, is this: pollution is lowering global sperm count levels. "Rigorous and comprehensive analysis finds that [sperm count] declined 52.4% between 1973 and 2011 among men from Western countries, with no evidence of a 'leveling off' in recent years."[34] A 2017 study, which was a collaborative effort by doctors at Hebrew University and NYC's Icahn School of Medicine, found that an increasing proportion of males have sperm counts below any given threshold for subfertility or infertility. "The high proportion of men from western countries with concentration below 40 million/ml is particularly concerning, given the evidence that SC below this threshold is associated with a decreased monthly probability of conception."

The public health implications are even wider, though. Poor sperm count is associated with overall morbidity and mortality—that is, it's a harbinger of other problems that are likely coming at us.[35] "In particular," the study continues, "endocrine disruption from chemical exposures or maternal smoking [and here we remember that in high concentrations air pollution can mimic or exceed smoking in its harm to those who endure it] during critical windows of male reproductive development may play a role in prenatal life, while lifestyle changes and exposure to pesticides may play a role in adult life. Thus, a decline in sperm count might be considered as a 'canary in the coal mine' for male health across the lifespan."[36]

Life expectancy and quality of life

The Centers for Disease Control found that American life expectancy dropped by 2.7 years between 2019 and 2022.[37] COVID-19 was certainly the central factor in these declines, but other top factors include: unintentional injuries (for instance, drug overdose deaths account for nearly half of all unintentional injury deaths), and chronic liver disease and cirrhosis for people of all genders, stroke and heart disease for women, and suicide and homicide for men.[38] This tracks with reports that Americans' health-related quality of life has been in decline since the turn of the century, as well as with our lived experiences of increased suffering and strife around us.[39]

And while we've focused here on reproductive harms, it should be clear that the environmental factors that increase the difficulty and risk of conception, gestation and infant survival also contribute to general ill-health: increases in respiratory and heart disease, cancers, lowering life expectancy and quality. We have reached the saturation point of fossil-fuel poisoning.

"Every disaster you have yet to experience in your lifetime has already begun," writes Samantha Montano, in *Disasterology*. "The threads of risk are spun out over decades, even centuries, until they crescendo into disaster."[40] Dr. Ha elaborates, "as climate conditions deteriorate, less healthy parents will be giving birth to less healthy children, resulting in a repeating cycle of unfavorable health outcomes that will impact generations to come."

Some of these outcomes are already set in motion, but mitigable. Some are as-yet unwritten, and so, preventable. Dr. Ha puts this in clear public health terms: "Although individuals can potentially minimize their own risk to a certain extent, the ultimate solution lies at the population level." Our job right now, as people organizing for a safer future, is twofold: stop the harm we can, and mitigate the harm we can't stop.

So Where Does This Leave Us?

Dr. Monteblanco declines to make recommendations about what would-be parents "should" do, or consider. "I will not add to the pressure, guilt, or overwhelming responsibility of parents," she says firmly, "especially because not everyone is in a position to *consider or choose* parenthood. I work to make the earth a little safer for pregnancy and childhood." The sad reality is that people planning families have to navigate dangers on all sides, at all stages of procreation, often with little or no institutional support. But to change that reality we can join growing movements intent on these twin jobs of stopping and mitigating harm.

"I do agree with the WHO's view that climate change is the biggest public health threat of the 21st century," says Dr. Ha. "It is an existential issue that can only be solved by concerted efforts that can be catalyzed by activism and advocacy, especially for groups that are more impacted (who often do not have a voice)." And Dr. Monteblanco agrees. "The maternal health crisis is absolutely a call to activism, advocacy, more research funds, shifts in training, new policy adaptations, and continued mobilization from everyone and an important reminder to offer leadership opportunities for women of color," she says. "And a call to pass federal policies like the Momnibus (including the Protecting Moms and Babies Against Climate Change Act) and the Pregnant Workers Fairness Act."

We'd like to conclude by returning to Yudith Nieto. We asked her what keeps her committed to the difficult work she does, and she says plainly: "I'm in so much awe of the resilience of the earth, of plants, of ecosystems." Despite contamination, flood, and drought, "it still continues to thrive. And I just keep watching it, and my spiritual commitment to earth and to myself, my body, my communities, you know, plant communities, human communities,

and animal communities. That's what keeps me committed."
We'll meet back up with Yudith in the second half of the book
as we delve into activist strategies. For now we're sharing her
hard-won insight that, as well as causing her to fight for herself
and her community at a young age, the fossil-fueled damages she
witnessed, and withstood, have shaped her hope, and her understanding of healing:

> I don't have grand visions of wealth other than my health. I
> want to be a healthy person. I want to come to the point where,
> when and if I want to have children of my own, I'm able to do
> it. I trust my body will hold that, and will hold me afterwards.
> If I'm able to stay healthy, able to acquire foods that are healthy,
> and able to cultivate the land in a healthier way, just like my
> ancestors had done before we came here, I can say I lived a very
> wealthy life. And if I am *not* healthy at some point, if I live the
> consequences of being exposed to so many hard carcinogens,
> I hope my body will be as resilient as the Earth has been. I've
> already dealt with asthma most of my life. And so to think
> that at some point, any of my family members or myself comes
> down with cancer, I want to believe that there is still some resil-
> iency on this earth to find healing again. That's what keeps me
> committed, to be able to continue to heal even if I'm not ever
> able to be well. I wish that for everyone.

CHAPTER 2

But Are Babies Bad for the World?

As we just saw, there are so many ways that the climate crisis is making it riskier, more toxic, and less equitable for people planning families. It's surprising, then, that these findings haven't been at the heart of the climate-and-babies conversation. But even more surprising is how thoroughly the public conversation is devoted to the false climate driver, and the false climate solution, of population.

In 2014 when we started Conceivable Future, on the rare occasions that climate and reproduction were discussed together, they were always framed the wrong way around: focusing on childbearing's impacts on the climate. As we began talking with people about their reproductive lives in a changing climate, we found that populationist rhetoric was a major obstacle to *just having this conversation*. In early media coverage we were often wrongly assumed to be populationists. The deeply ingrained population myth has been pervasive in the Global North for so long that it is now largely understood as common sense.

To understand more about how these issues came to be so tightly knotted together, we spoke with Dr. Jade Sasser, a professor in the Department of Gender and Sexuality at University of California, Riverside. She explains that this narrative's ubiquity comes in part from the work of the organization Zero Population

Growth (now called Population Connection) who did "a lot of work to develop school curricula, and to distribute them to educators across the country who were working at the elementary and junior high levels. And their goal was broad dissemination of messaging about population and its environmental effects, so that every school child in the public school system would grow up learning about the destructive impact of population from an environmental perspective."

In its milder expression, this conventional wisdom might take the form of anxiety about the emissions that children "cause." We've heard this often from participants at our events. In other parts of environmental discourse—and in our inbox, as we'll get to shortly—this can take more extreme and antisocial forms.

The argument for reducing the human population as a way to reduce emissions feels compelling in its simplicity: more people means fewer resources, more hunger, more suffering, more pollution. By reproducing, poor people create their own poverty. By focusing on population size, Global North institutions, whether governments or NGOs, are absolved from both responsibility and nuance!

Through our organizing we set out to understand a complex set of concerns. We did not set out to shape, or even address, the size of the global population. But we quickly found that in order to have our conversation, we needed to understand the history and ideologies of population movements, as well as the global feminist countermovements for full reproductive autonomy and against population control.

Shedding light on the history—and present-day reality—of population control efforts has been a necessary part of this project from those early days. The history of population control is a transnational history of powerful people deciding whose babies are desirable, and whose babies are to blame for any number of societal ills. Indeed, as Dr. Sasser explains, before it came to be widely associated with environmental problems, population was socially constructed "through the efforts of people who were deeply concerned with managing a range of issues: 'socially undesirable' traits like criminal tendencies, alcoholism, mental illnesses, cognitive delays, unwed pregnancies, and dependence on welfare, global geopolitical instability and armed conflict; immigration; economic development in newly independent former colonies,"[1] as well as the spread of communism in the Global South. The solution to any social problem, when viewed through populationism, is not to fix it but to make it disappear *by eliminating future people.*

Thomas Robert Malthus, a cleric and economist (and one of seven siblings) laid the foundation of this bad ideology in *An Essay on the Principle of Population*, published in 1798. And his claim that "misery" was a natural result of unchecked human reproduction, rather than a human sacrifice to capitalist exploitation and resource extraction, prefigured the neoliberal attitudes that keep population-focused projects alive today. Gone are many of the human-rights violations, and the more egregious language, but intact is the assumption that it's better to pay the costs of capitalism by regulating fertility than regulating markets. Much of this work is international: aid and development organizations from the Global North instrumentalize family planning and contraception distribution toward various environmental and economic aims in the Global South. But the same philosophy and practices play out on American soil, following familiar class-based and racialized contours.

Not only does this approach obscure real solutions, it identifies the wrong problem. A 2012 study of global economic drivers of carbon emissions found "no relation between short-term growth of world population and CO_2 concentrations."[2] As Dr. Sasser points out, the closest correlation to a country's emissions is their Gross Domestic Product.[3]

It's important for us to note here that there has long been scholarship and activism that debunks populationism; we're connecting some dots, not breaking new ground with this critique.

But we find that the critique itself is still mostly absent from mainstream environmentalist spaces, and from the climate movement specifically—where it's needed most. As we've learned more about this legacy, we've become increasingly conscious of the subtle and persistent forms of population ideology in even progressive climate activism. On the personal level this can manifest in activists who reject overt populationist policies but still feel guilty about their own desire to bear a child, who sit in judgment of any larger-than-replacement-rate family, or who believe uncritically that adoption is more ethical than biological reproduction.

In this chapter we address the history and biases that are shaping people's views, and from this perspective look critically at calls for individual behavior-based solutions. We also look to the Reproductive Justice movement—its history and analysis—as an essential corrective to the climate "commonsense" approach to population numbers. This is not just a general, moral point, but a time-sensitive and practical one: the climate crisis has emboldened populationists with new urgency. Human rights, women's rights, the right to bodily autonomy, have always been imperfectly—and unequally—defended. Now what used to look like slow progress is more like holding the line, or backsliding. Dr. Sasser points out that after the 2022 *Dobbs* decision, which ended the right to guaranteed abortion access at the federal level in the United States,

> one of the direct effects of that could be that it would actually increase human numbers in the United States, more people being born. Was that the intention? Who knows? But there has been a lot of evidence that came out in around 2020–21, that demonstrates that population and fertility rates in the US have been falling for about 80 years now. And demographers have known that. . . . In some quarters in the US, there is a lot of concern about that: what the impacts will be on the economy, on the society as a whole, etc. That's very different from the continued focus on population growth outside of the United

States. Particularly in global South countries, poor countries, places where the livability of a particular community is under threat in a much more immediate way.

Those of us in the United States are living in a dangerous confusion of policies that both push and pull at all our rights to reproductive self-determination. The narrative is not as simple as "have more babies" or "have fewer babies." Rather, it is: "Your body is not your own." Today more than ever, beware of population "solutions," which are at best ineffective, instrumentalizing, and freighted by white supremacy and classism. Whatever problem it names, the population "solution" punches down, enabling powerful players to evade responsibility while continuing to harm.

As we have learned from justice movements—and faith traditions—the path is the goal. The values we hold dear, of self and community sovereignty, anti-fascism, and human rights, are what we enact to find our way to the world we want to live in. There are no shortcuts through gray areas.

Someone Else's Babies: Problems with the Population Conversation

Look in the Conceivable Future inbox and you'll see a folder that Josephine, in a moment of inspiration, labeled "barf." In it you'll find some of the worst emails we've gotten over the years; trolls and bullies, mostly. But one notable genre of unsolicited communication is a kind of amateur policy paper from aspiring authoritarians about how to control the global population. The authors tend to be from the United States, Canada, or Western Europe, mostly (but not exclusively) men, and although they don't always share their demographic information, they tend to be white and of retirement age. Their ten-point plans will propose some combination of carrot-and-stick strategies for reducing births, and they typically want us to elevate their plan. Maybe we could send it on to the UN, or share it through our socials?

Some other features these strange missives have in common: the authors think they've arrived at an ignored—or unjustly repressed—solution; they think it's the silver bullet; they think we are an anti-natalist group and therefore simpatico to their program; they're touting the *affordability* of this as opposed to other more popular climate initiatives.

We put these and other letters in the barf box for a few reasons. First, because these individuals fundamentally misunderstand what we're trying to accomplish, but we save them because they tell us about what we're dealing with. Second, as should be clear by now, we find these letters deeply disturbing. And third, we have to make some horrors into a joke or we'd never get out of our beds, let alone open the inbox.

We will never achieve a more just world by curtailing people's reproduction. We don't get there by control, coercion, or force. The future we are struggling for stands on a foundation of human rights, in which we share and defend full self-sovereignty.

The argument for population control is based on three interwoven and equally toxic assumptions: first, that rapid population growth is the cause of "underdevelopment" in the Global South; second, that policy should persuade (or if necessary, force) people to have fewer children rather than improving the conditions in which they live, and; third, that some combination of finance, managerial, technological, and Western intervention techniques can "deliver" birth control to the Global South in a top-down fashion *in the absence of comprehensive healthcare.*[4] In the priority promotion of contraceptives, the premise is clearly that pregnancy prevention matters most.

These days population-focused international aid work isn't synonymous with human rights violations. The ongoing problem is in the priorities. Dr. Sasser learned this firsthand as a Peace Corps volunteer in Madagascar, and later as an NGO worker focused on family planning. "It's not like I am suggesting that these institutions are sending people into poor countries to coercively round people up and sterilize them or anything like that," she says. "It's a narrative. But that narrative makes it possible to fund particular kinds of work. And where that work is problematic is, for example, in places where, let's say, girls and women have all sorts of reproductive challenges, and health issues, and health concerns. But when they go to the clinic, the only kind of reproductive services they can access are family planning services and maybe STI prevention."

Population scholar and author Betsy Hartmann describes how contradictory political views have converged on this singular issue: "The early Neo-Malthusians supported birth control as a means of improving the condition of the poor by limiting population growth; feminists and socialists believed it was a fundamental woman's right; eugenicists embraced it as a way of influencing genetic quality. These strange bedfellows combined to give the birth control movement its unique character: it carried within it the seeds of birth control as a liberating force, as well as a means of coercive population control."[5]

Although this thinking developed as part of an international development agenda, countries have engaged this ideology domestically, plenty. In the United States alone, the government, as well as NGOs, have repressed the reproductive rights of Indigenous women and women of color by implicit and explicit policies for hundreds of years, from the foundational policies of genocide, slavery, and abuse to the present day. Between 1909 and 1979 approximately 20,000 people were forcibly sterilized in California, a practice that continues to this day in Canada and elsewhere.[6] In the United States, the idea of the IUD as a "cure" for poverty has a disturbing hold on public discourse, and as recently as 2017, judges have shaved off sentence time for people who agree to undergo sterilization procedures.[7] Black women are still disproportionately targeted for violent reproductive interventions, including abortion, sterilization, and contraception.[8]

These practices are rooted in a lethal trifecta of sexism, racism, and classism. Cultural attitudes about population amplify beliefs about who is a "fit" mother.[9] And the concept of "fitness," in turn, derives from ideas about who is or is not a valuable human being, or (in the eugenicist tradition) what are "valuable" traits and genes. Around the world we find legacies of reproductive violence and oppression.

The Magdalene Laundries of Ireland are one notorious example. The Laundries were part of an interlocking system of orphanages, industrial schools, "mother and baby homes" for unwed mothers, and church-run institutions in which thousands of Irish citizens were once confined. Roman Catholic orders of nuns ran the (for-profit) laundries, and women and girls were made to work there, nominally as a form of penance for their sins. The laundries were filled not only with "fallen women"—prostitutes, women who became pregnant without being married or as a result of sexual abuse—but also those who simply failed to conform to the expectations of their society. Children born to women in the laundries had their names changed and were adopted out without their mothers' consent.

As another example: during the early years of the Salvadoran Civil War, which lasted from 1979 to 1992, the military, which had led a coup, took thousands of children from their (anti-coup) families. The children's identities were changed and they were sent abroad for adoption, primarily to the United States and Europe. In other words, El Salvador also "disappeared" children of people who were identified as insurgents, *as a way of controlling adults*. These operations were carried out by lawyers with military contacts and foreign adoption centers, that watchdogs have since flagged as part of the international human trafficking black market.

As recently as May 2017, a Tennessee judge issued a standing order allowing inmates to receive thirty days' jail credit in exchange for undergoing a voluntary—for the dubious value of voluntarism while incarcerated—sterilization procedure.[10] The message here was unmistakable: that people who wind up in prison should be bribed to rescind their human right to have children.

As all three of these examples show, when powerful people make moves to determine who is "fit" to be bearing and raising children, these determinations target marginalized people, and women specifically, marginalized communities more broadly, and often reward families from the dominant group, *sometimes even with the children of those marginalized groups.* Even when those biases are not an explicit policy, wealthy people are more apt to be considered fit parents. These practices have been a weapon used by the powerful to control the less powerful. This is the history of population movements—a history we must learn, and whose wounds are still open.

This kind of paternalistic repression is having a renaissance in the context of the climate crisis. Our movement's recent history shows us that when our societal focus shifts to policing the behavior of private citizens, we can be handily distracted from the crimes being committed right over our heads. Confronted with an urgent need for change, who do we pressure for that change: those with the most or the least power? One of those projects is certainly *easier* than the other.

And since its early days, the mainstream environmental movement has recruited these strategies in service of "conservation" goals. For instance, the best-selling book *The Population Bomb*, written by Paul and Anne Ehrlich and published in 1968 at the suggestion of then-director of the Sierra Club David Brower, predicted worldwide famine in the 1970s and 1980s due to overpopulation. It also prophesied other major upheavals, and advocated immediate action to limit population growth (suggestions in the book included the idea of adding birth control to the food or water supplies). The Sierra Club—one of the best-known environmental organizations then, and now—actually sponsored publication of the book, and during the 1980s some members (including Anne Ehrlich) steered the group into the field of US immigration, arguing that overpopulation was a significant factor in environmental degradation, and advocating halting and reducing US and world populations. In 1988, the organization's Population Committee and Conservation Coordinating Committee argued publicly that immigration to the United States should be limited, so as to achieve population stabilization.[11]

When population control and border control efforts converge, it's easy to see that both programs are built to uphold the inequality of the status quo. And even when the tactics are not explicitly violent, the paternalistic disregard for people's rights is plainly evident. Overpopulation is an ideology; it diverts criticisms of capitalist consumption and unequal distribution by blaming devalued people—mostly women, and often poor women—for reproducing. It serves to justify a system (capitalism) that creates needs among many while satisfying them only for the very few.

The most notorious and widespread campaign to control population size was China's One Child policy, which began in 1979 and continued for more than three decades. In some environmental corners this policy is even quietly admired; by the Chinese government's projections, its population would have been 1.8 billion without it, instead of today's 1.435 billion. Some scholars dispute this claim, arguing that "as much as three-quarters of the decline in fertility since 1970 occurred before the launching of the one-child policy," and that "most of the further decline in fertility since 1980 can be attributed to economic development, not coercive enforcement of birth limits."[12] Aside from the policy's matter-of-fact disregard for citizens' human rights, it also caused major societal problems, including the proliferation of sex-selective abortions and the resulting scarcity of adult women, and the emotional scars that people continue to carry after forced abortions, sterilizations, and massive fines and jail sentences for violating the law.

And while this was happening, China's emissions rose to surpass those of the United States. Turns out that human rights abuses aren't the key to decarbonization, after all.

To its credit, the Sierra Club has reversed many of its organizational positions on population, and done so quite publicly (although there are still groups within the Sierra Club advocating for a return to a population orientation, and the controversy resurfaced when three anti-immigration proponents ran in the 2004 Sierra Club Board of Directors election).[13,14] But these harmful ideas about population are still firmly lodged in the public consciousness; ideas that logically manifest in racist, xenophobic, and violent ways. The men who perpetrated mass shootings in Christchurch, New Zealand, and El Paso, Texas, cited "overpopulation" as a reason they targeted immigrants, many of whom are fleeing the devastating effects of climate change in their home countries. It is a short step from viewing "overpopulation" as a problem to any number of violent "solutions."

One of the most painful ironies of these ambitions to control women's reproduction is that they're functionally unnecessary when the US government offers so little support to women when they *actually become* mothers. In other words, what all of the emails in the barf box overlook is that the United States is already perfecting anti-natalist policies in all but name. As we explored in more detail in the last chapter, the United States is the least welcoming place out of all the wealthy countries for new parents. Asthma and other diseases are on the rise from deregulated air pollution, fracking chemicals, and other industry contaminants. Healthcare costs are staggering. An uncomplicated hospital birth costs $32,093 on average.[15] Maternal mortality is higher in the States than it is among peer countries, criminally so for women of color. Food deserts abound in our cities, public schools are shuttering or starving for funds, and daycare costs more than college in many places, never mind the costs of college itself.[16]

Even bracketing the climate crisis, anyone considering a family in this place, at this point in time, is already assuming a burden of medical and financial risk. And each of these factors that weigh against people's reproductive lives are doubly weighted against people of color. As Dr. Sasser found when she interviewed American women aged twenty to forty about their feelings toward climate and reproduction, her subjects' emotional experiences were strongly conditioned by race. She spoke with women from across the racial spectrum, and she found a high concordance among all participants' concerns. But for women of color,

> the concerns about climate change, the concerns about mental health issues were heavily compounded by experiences of racism and perceptions, and experiences of racial vulnerability. Meaning: the women of color that I interviewed strongly perceived that their children would already be saddled by issues of inequality just for existing as Black or brown, or indigenous. So that they would have to fight to ensure that they had quality education in school. That their children would at some point potentially have to face the criminal justice system or deal with police, police brutality, police violence, or just being treated differently by police. They knew that their children at some point would have to deal with some kind of racial discrimination that would be very hurtful to them.

As we mentioned before, most of the population-control epistles in our barf box have come from older white men who usually have a child or two, maybe grandchildren of their own. They tend toward the egalitarian edge of a broader trend because they include white, middle-class American women in the population they intend to control. But the trend is this: they find it easier to imagine reshaping young people's reproductive lives than to even imagine reshaping parts of the economy. They have had children, but they are here to tell us that *we* shouldn't, for our own good; they made no sacrifice themselves, but they are writing to demand it of their young, while blithely ignoring the toxicity, injustice, and lack of support that already inhibits American reproductive freedoms.

In other words, young people in the United States don't need any more disincentives to have families—we already live in a country that is outright hostile to parents and children. And indeed, the US birth rate hit a thirty-two-year low in 2018, with millennials reporting that they are having 1.5 fewer children than they'd like to have, on average.[17] And the COVID-19 pandemic exacerbated what was already there; birth rates plummeted in wealthy countries, including in the United States.[18]

This shows us two things: first, many climate-minded people badly need to learn about the reproductive realities in this country. And second, that even people with a wrongheaded but sincere concern about climate change (the writers of these proposals, for instance) seem to think it's easier to tell a whole generation what to do with our bodies than to put the fossil-fuel industry on notice.

Present-day population advocates are eager to put distance between the "dark past" and present-day "empowerment"-focused family planning programs. And indeed, voluntary access to birth control is a marked improvement from crimes of coercion and violence. But even in its most empowering form there are, to us, three major and connected flaws in the arguments for smaller populations as a "climate fix."

The first is that these arguments provide cover for eugenicist dogma, whether we consciously espouse it or not. *Women in India are having too many babies. New Hampshire trailer trash doesn't know how to use birth control.* These accusations—real-life examples—are both racist and classist; how many times have you heard a middle-class white woman accused of having "too many babies?" In fact, in the early days of Conceivable Future organizing, we—two white women—were frequently told by white observers, "but you're the people who *should* be having children."

(This was a particular irony because Meghan was not always a middle-class person and is from a relatively large family herself. She moved from being told she *shouldn't* have babies to being told that she *should*, as she moved into the middle class.) We encountered another manifestation of this bias when at several house parties we met white middle-class women, confident they never wanted to become pregnant, who couldn't find a doctor that would perform a voluntary sterilization on them. The difficulties some women face in getting a sterilization procedure as a form of contraception are long-standing and well-documented.[19] In this country's context of involuntary sterilizations for BIPOC women and incarcerated people, the irony is ghoulish. It's important to note that "overpopulation" is a term used overwhelmingly to describe the demographics of poor areas and/or nations. In punditry about rich countries, we're much more likely to hear about "underpopulation" and its purported negative effects on the capitalist economy.

The argument's second flaw is that it offers convenient scapegoats for systemic overconsumption in the rich parts of the world. *Population relates to climate harm only to the degree that populations exploit resources and emit carbon.* No one has emitted more than Americans. It's not the number of people alone; arguments about "optimum population" (such as those that Ehrlich is still making) ignore or minimize the systemic nature of resource consumption.[20] Waste is a feature, not a bug in our industrialized systems: look to planned obsolescence, low-gas-mileage vehicles, and the excesses of conventional agriculture: a recent study claimed that every other bite of food in the United States is waste.[21] Nothing about being a human requires this, nor does it correlate with happiness or a high quality of life.

To bring this home we've often said that if everyone on earth consumed the way the American upper and middle classes consume, we would need an additional 4.5–6 Earths' worth of resources to sustain ourselves—a fact absent from much populationist rhetoric. But in early 2023 a study published in *Ecological Economics* broke the population argument down even more finely along consumption inequality lines. We asked the authors of this study to help us put it in plain terms, and Dr. Jared Starr obliged in an email:

> The Global Footprint Network estimates that if everyone consumed like the average American we would need 5.1 Earths (in 2018—the latest year I found data for). Carbon emissions alone account for 3.65 Earths. We find that average top 0.1% US households have emissions 23x higher (954 tons) than the US average.[22] Multiplying how many Earths are needed for average Americans' carbon emissions (3.65) times 23 I estimate that *if everyone on the planet emitted carbon like an average top 0.1% US household we would need 84 planet Earths.*

The data gets richer. Dr. Starr pointed us to a study about carbon emissions from twenty billionaires, which found that average emissions of those people were 194 times higher than an average American household.[23] Using these numbers, Dr. Starr told us, "if everyone on the planet emitted carbon like a billionaire, we would need something like 700 planet Earths."

These numbers certainly put a fine point on where the culpability for carbon emissions lies. (Spoiler: it *isn't* on the shoulders of women seeking affordable healthcare from international aid organizations.)

So let us be perfectly clear: we need policies that balance a global standard for quality of life with rapid decarbonization, and with a progressive focus on dematerialization for the West. No valid policy involves repressing human rights or outsourcing responsibility.

The other big problem with the population climate "fix" is that it instrumentalizes women's bodies and our healthcare, especially in the Global South. In other words, this perspective assumes that women should be able to access healthcare, contraception, and education *because* those things support the goals of decarbonization. And it's worse than that: the real point is that if the goal of family planning services is to reduce population growth—rather than to support the freedom of people to determine the number and spacing of their own children—women can expect inferior care. Earlier Dr. Sasser described how the narrow overfocus on contraception she witnessed as a Peace Corps volunteer in Madagascar was not serving the sexual and reproductive health of the people she intended to help.

When she returned to Madagascar years later on a research trip for her dissertation, she had conversations with friends working in USAID, who were overseeing the funding of clinics throughout the entire country, that illuminated the central conflict of interest. She asked her friends how USAID was serving reproductive health clinics that might be helpful to the community she served: "'What are you doing that I could maybe bring back to this town that I had lived in to better help girls avoid pregnancy?' And I discovered that there was a strategic plan as to where these services would be prioritized, the reproductive health services that were funded by USAID. And it was all around national parks and conservation sites."

The implicit priority here was to keep population numbers low to preserve pristine landscapes and rare animals, and, presumably, the revenue that these places generate as tourist destinations. In this view, more pregnancies would lead to more demand for land, water, firewood, and so forth, and perhaps eventually more poachers. Dr. Sasser continues, "I was like, *What is going on here?!?* There are not more girls getting pregnant near parks and conservation sites. Why would your environmental agenda be the agenda that organizes and determines the population reproductive health and family planning agenda? Honestly, if there is one moment that I can point to where a light switch flipped on for me, that was it."

And this was not an isolated situation of conservation policy determining the kind of care women received. In 2019, Dr. Sasser and a colleague copublished an article "about the services that are offered to people in health clinics [and] reproductive clinics in Madagascar, in areas that are near marine conservation sites." Her colleague who did the fieldwork discovered, "over, and over again that even when women came into clinics for other concerns, with other needs, wanting other services, they were consistently steered toward contraceptives." That is: first, these clinics essentially defined "women's health" as their capacity to get pregnant, and second, they used birth control for their own conservation goals, rather than any health goals belonging to the women themselves.

In the popular climate solutions handbook *Drawdown*, Educating Girls and Family Planning are ranked as #6 and #7, valued at a combined 51.48 gigaton reduction of CO_2. The author clarifies that "when family planning focuses on healthcare provision and meeting women's expressed needs, empowerment, equality, and wellbeing are the result; the benefits to the planet are side effects." Even so, those side effects are the whole point of the book. Access to education and family planning are human rights, and they are rights that women deserve because we are human beings. By treating access to those rights as a means to carbon reduction—or habitat protection—rather than an end unto itself, climate groups continue to behave toward women as though we are second-class citizens, or simply valves to be turned.

In fact, a recent study shows that conformity to masculine stereotypes correlates with environmentally harmful behaviors.[24] "Caring about the environment" is widely seen as a feminine set of behaviors, so a person concerned with appearing masculine is measurably less likely to recycle, value fuel efficiency when purchasing a car, and so on. And it's not just men upholding a bizarrely fossil-fueled masculine ideal: Georgia Representative Marjorie Taylor-Green claimed that "Democrats like Pete Buttigieg want to emasculate the way we drive and force all of you to rely on electric vehicles."[25]

We are eager for the debut of the Drawdown program's "Educating Boys and Detoxifying Masculinity" targets.

This kind of discourse—the "let's support women's healthcare only as a means to carbon-reduction" discourse—exists in the context of a culture that's generally only interested in women's health insofar as it serves some other purpose. The medical establishment has a well-documented male bias and a tragically consistent history of ignoring women's healthcare needs.[26] Dr. Kate Young, a public health researcher at Monash University in Australia, has found that women are often viewed by their doctors as "reproductive bodies with hysterical tendencies"—an orientation from which flow any number of distorted outcomes. Medical experts routinely dismiss women's healthcare complaints as invented or psychological, making comments like "There's a lot of psychology, just as much as there is pathology [in gynecology]."[27] One doctor claimed that he'd never met a fibromyalgia patient who wasn't "batshit crazy."[28]

And these experiences—while present for many women all over the world—are especially present for Black women in the United States. In a now well-known story, tennis superstar Serena Williams had to bring the full force of her stardom to bear before her postpartum blood clot was taken seriously by medical staff.[29] "When you are a Black woman, having a body is already complicated for workplace politics," writes Tressie McMillan Cottom in *Time* magazine. "Having a bleeding, distended body is especially egregious." The medical establishment filters Black women through assumptions of incompetence. "When the medical profession systematically denies the existence of black women's pain, underdiagnoses our pain, refuses to alleviate or treat our pain, healthcare marks us as incompetent bureaucratic subjects. Then it serves us accordingly," she concludes.[30]

The point here is that alternately ignoring and instrumentalizing female bodies is a long tradition that is shaped and torqued by race, class, and geography. The current discourse of women's-health-qua-climate-solution fits neatly within a long-standing patriarchal tendency to view women as simply "reproductive bodies with hysterical tendencies" and to treat us as such. That is: women, and our experience of our health, are routinely dismissed and ignored, *except* when someone gets the idea that managing our fertility in some way would be good—and let's be clear, *cost-effective*—for some other social agenda. Other such agendas have included: populating the Fatherland, producing more people to enslave, and preventing the reproduction of those deemed less worthy.

At this moment in time, the desired outcome is cheap and "easy" carbon reduction.

Here is the bedrock belief from which we challenge these views: women are human beings, and we deserve healthcare, which includes full-spectrum reproductive care, *because* we are human beings. Not because many of us have uteruses, or because those uteruses could be requisitioned to produce more—or fewer—people.

One's Own Babies: Problems with the Lifestyle Conversation

In 2017, Lund University in Sweden published a widely circulated study, coverage of which touted four "lifestyle choices" to reduce your carbon footprint.[31] Topping this list was to have one fewer child. The study's lead author, Seth Wynes, said, "Those of us who want to step forward on climate need to know how our actions can have the greatest possible impact."

While one could argue that children are a "lifestyle choice," thinking about it this way trivializes the existential significance of this question for many people. Having one fewer child is not akin to adopting a bike commute or a vegan diet; it certainly shapes one's identity, social connections, and life experience in ways that transport and food do not.

The Lund paper uses data from a 2009 study that calculates a parent's carbon footprint by attributing half of an average westernized lifetime of CO_2 emissions to each parent, plus a quarter for a grandchild, an eighth of a great-grandchild, and so forth.[32] When figured this way, the numbers are impressively large, for sure. But what they're implicitly saying when they base linear projections on today's rates of consumption is this: "We will not change the systems that lock us into a high-carbon way of life."

In other words, the study assumes that we have no capacity to change the conditions under which we live that make parenting so carbon-intensive in the first place.

We already know that Jeff Bezos and his peer group's unacknowledged presence in this calculation massively bloats the consumption average. But this study is premised on many more layers of flawed thinking. First, if we believe that per capita consumption will not decline in our children's lifetimes, then focusing on individual family size is an unworkable solution anyway: reproductive decisions take effect on a slower time frame than the closing window for decarbonization. Babies' emissions live in their future adulthood and beyond: the carbon cost of someone born today won't be paid in full until well after that window is closed. Even in a *Children of Men* scenario, in which zero babies are born starting today, emissions wouldn't decline until a substantial portion of the aging population had died. And whether by our concerted action, our failure to avoid collapse, or the likelier combined effects of both, *none* of today's babies will ever consume as much as we, their parents, do now.

Just as striking as what the study (and the subsequent punditry about it) includes is what it *doesn't* include: any conversations about organizing for change. The framing of one-fewer-child ignores the systemic origins of the climate crisis, and seems to suggest that we could solve the problem by just . . . not participating in the system. This individualist narrative has driven and obstructed the climate solutions conversation for years. It sends the message that climate change could really be fixed if we all just *behave* better: use fewer straws, drive a Prius, and—crucially—have fewer children. It belies a belief that the industrial and existential problem that is climate change could be "fixed" if we all dial down our consumption. But, if we've said it once we've said it a thousand times: this is a very dangerous attitude. It is also a woefully inefficient theory of change: individual people's decisions are the smallest levers available, that are then operated only through mass emotional manipulation. The emotional strategy turns both the climate crisis and our response to it into an individual choice, which is then defined as sin or virtue. That is an alienating, guilt-producing orientation that would at best reduce emissions slightly, while deterring people from political action.

But more importantly, this attitude deflects attention from the real perpetrators: the fossil-fuel industry, and the economics and politics that enable it.

And while we clearly believe it's necessary to consider one's family decisions within the context of climate change, we object to this weird move: making people's reproductive lives into the site of climate "action." If we treat having fewer children as the most significant action we can take—an idea that fits squarely within the neoliberal rhetoric of conscientious consumerism—we affirm the lie that the climate crisis is primarily an individual burden to be borne by individual households, and we evade our responsibility to organize and demand solutions that are big enough, and fast enough.

In other words, a consumption-oriented perspective has real limits. It cannot save us. Environmentally thoughtful choices (ranging from reusable tote bags to LEDs, carpools, and cloth diapers) are obviously good. And while personally both Meghan and Josephine endeavor to live thoughtful lives, using our resources responsibly, we are both implicated in this system. That system is bigger than each of us, bigger than our individual choices, bigger than any babies we, or you, have or might have.

Babies aren't the problem. Having one fewer child would not be the "biggest lifestyle change" one could make to reduce carbon emissions if the systems that governed our lifestyles were less broken. Our job is to fix the system, rather than blaming babies who aren't even born yet for consumption that is enabled—and indeed required—by the economy we live in.

REPRODUCTIVE JUSTICE AND A WAY FORWARD

In the previous chapter, Dr. Adelle Monteblanco named reproductive justice as the vision that guides her public health work. She quoted SisterSong, the largest multiethnic reproductive justice collaborative in the country, which defines reproductive justice as the human right to "maintain personal bodily autonomy, have children, not have children, and parent the children we have in safe and thriving communities."

Reproductive justice encompasses "reproductive rights" (legal access to reproductive healthcare, abortion, and contraception), as well as the economic and social dimensions of well-being that affect a person's sexual and reproductive life, as well as questions of access. The last category includes things like: accessibility of those reproductive services (how many state lines do you have to cross to get an abortion, and how much does it cost?), prenatal and pregnancy care (are these services covered by insurers? How much does giving birth cost?), domestic violence prevention, safe housing, healthy food, and adequate wages, among others. This is the inclusive political understanding of reproduction we urgently need in the climate movement.

The National Black Women's Health Project, formed in 1984 (and renamed the Black Women's Health Imperative in 2003), was among the first such organizations in the United States, and the foremother of many others, SisterSong included. Among its founders was Byllye Avery, a Floridian who was drawn to this work at a time when Black women were dying at a staggering rate from illegal and back-alley abortions. Avery's cofounder, Lillie

Allen, led pioneering group explorations of internalized racism. The conference that ultimately birthed the National Black Women's Health Project was based on a model of collective leadership, meaning that everyone involved was equally responsible for all the decisions (which Conceivable Future practices in our own organization). The model was deliberately designed to counter the pervasive charismatic model that the Civil Rights Movement employed, which rallied behind a small number of dominant male figures. The conference was thoughtfully organized and well-attended, and conference participants demanded an independent organization that was itself dedicated to Black women's health. The National Black Women's Health Project came out of this moment, and from its inception, its founders understood and explored the connections between racism, exclusion from power, and poor health. As such, its organizers understood that they would need to address racism, poverty, self-esteem, and extreme stress—and these issues have driven the organization's agenda ever since.[33]

Although the NBWHP laid much of the groundwork for the emergence of reproductive justice, many of the founding RJ documents and alliances were developed in the lead-up to the 1994 UN-sponsored International Conference on Population And Development (ICPD), where an international coalition of feminists opposed population control, calling for human rights–based policies.[34] American feminists of color were connecting their domestic experiences of repression and exclusion with the same abuses taking place internationally through policies of population control.

On August 16, over eight hundred signatories, under the banner of *Women of African Descent for Health Care/Protecting Our Reproductive Freedom Coalition*, published a letter to Congress in the *Washington Post* calling for "a health care plan that provides universal coverage and equal access; comprehensive benefits which cover the full range of reproductive health care services, including abortion; and provides for the inclusion of Black women on national, state, and local planning, review and decision-making bodies regarding health care."[35]

Second Wave feminist reproductive activism in the United States had primarily focused on efforts to both secure and defend the right to legal abortion. These efforts have always been crucially important as the right to abortion is always contested, and is even less accessible today than it was ten years ago. Indeed the *Washington Post* letter reserves its greatest emphasis to declare "WE WILL NOT ENDORSE A HEALTH CARE REFORM SYSTEM THAT DOES NOT COVER THE FULL RANGE OF REPRODUCTIVE SERVICES FOR ALL WOMEN— INCLUDING ABORTION."

But the struggle for reproductive *rights*, framed as the right to "choose," is a story primarily of white women's priorities. As we've seen in this chapter, for Black women the struggle for bodily autonomy and self-determination has long included the right to have, as well as not have, pregnancies and children. (We note that here we are using the language of the foundational documents, rather than the more trans-inclusive language that has followed as the reproductive justice coalition has grown.) The reproductive justice movement now explicitly champions the rights of queer and trans people and all people of color, regardless of their documentation status. Still in 1994, thirteen of the activists and scholars who had organized and written the *Washington Post* letter, formed Women of African Descent for Reproductive Justice and created the conceptual framework named reproductive justice.

Before long, other groups who had been omitted or harmed by the conventional framing of "reproductive choice" and "reproductive rights" found common cause with Black women's health collectives. In 1997, members of Women of African Descent for Reproductive Justice, together with organizations from many different communities of color, founded SisterSong, which continues to lead the national movement, and shape its concepts. We look to these concepts, not just for a rights-based corrective to the oppressive view of climate, health, and population, but as a way to imagine climate solutions that *actually* make people's lives better.

And while we've been enthusiastic proponents of this framework, Dr. Sasser cautions that "different groups use the justice framing in ways that are really beneficial for their organizations and for their work, but that may not actually be in line with the groups, the activists, that created those frameworks to begin with. And I recognize people use this phrase really differently. But I am wedded to the original development and usage of the term, which is that reproductive justice is a response to histories of reproductive racism."

As the climate crisis—and the American war on reproductive health—intensifies, more people with racial and economic privilege are experiencing hardships, threats, and worries that are new to them. Reproductive justice benefits us all in that it offers a way to see how far-reaching the issue of reproduction is, and how many aspects of society are actually involved in supporting or undermining people's reproductive lives, health, families, and self-determination.

If we are able to communicate anything to other white climate activists, and/or to majority-white climate organizations, it is these two things. First, the reproductive justice framework can be transformative in understanding not just the population discourse of climate change, but also the crisis as a whole: from problems to solutions. And second, to make use of this framework, white climate organizers also assume a duty of gratitude to the movement that built it, through an explicit commitment to racial justice. Everyone has cause to demand a future with them and their families in it, but there is no climate justice without reckoning with reproductive racism, past and present.

The social movement landscape across the country right now includes many priorities that support reproductive justice (including housing, access to medical care, eradicating food deserts, and many other issues). The breadth of the work being done showcases just how expansive the necessary work is, and how it arises from within communities, rather than descending from policy in the abstract. If the framing of reproductive justice helps put a fine point on the problems, it can also help us define the work needed to solve the problems.

We'd like to end this chapter by looking at some of the work to enshrine principles of reproductive justice in institutions and laws.

One shining example is the increase of midwives and doulas across the United States, particularly in Black and Indigenous communities. As you now know, the maternal mortality rate in the United States is among the highest of developed countries. And that mortality is also highly racialized—Black babies die at twice the rate of white, Asian, or Latinx babies, pointing to a particular anti-Black racism enshrined in US institutions, including our hospitals and medical schools.[36] In response, communities across the country are turning to the support of local midwife collectives, such as Southern Birth Justice Network's Mobile Midwife Clinic in Florida, and Changing Woman Initiative's Corn Mother Easy Access Women's Clinic in Albuquerque, NM, and doulas.

Doulas of all races and in all contexts advocate for people giving birth, by offering a mix of education, advocacy, and emotional support. Given the staggering racial discrepancies in maternal mortality, Black doulas, who are trained in culturally sensitive advocacy, can be literally the difference between life and death. The emergence of Black doula collectives points to ways in which communities are creating their own solutions for well-being. Increasingly, grassroots campaigns to ensure that doula services are covered by insurance are meeting with success. As of this writing, California, Nevada, Illinois, Arizona, Rhode Island, Louisiana, Washington, Oregon, Minnesota, New Jersey, and Indiana have all passed bills relating to Medicaid coverage for doula care.[37]

The national push for universal healthcare, including full-spectrum reproductive healthcare, is another campaign that aims specifically at increasing healthcare access and reducing disparities along class and racial lines. Advocates of this approach—sometimes envisioned as Medicare for All—point out that astronomical healthcare costs and lack of access continue to drive people into bankruptcy and despair while many basic healthcare needs go unmet and insurance companies sop up billions of dollars. The vision is a single, national healthcare system for everyone in the United States, that would cover every medically necessary service, from routine doctor visits to surgery to mental health to prescription drugs, along with dental and vision care. Proponents argue that a universal, single-payer health care system would decrease health disparities, control costs, and grant everyone equal access to an excellent standard of care.

A third example, which has become critically important in the aftermath of the *Dobbs* decision that overturned federal abortion protections, has to do with legal protections for abortion care itself. Efforts are underway nationwide to ensure legal protections for abortion at the state level; Kansas voters robustly upheld the right to access safe, legal abortion in a 2022 referendum. Campaigns like this address the issue of legal rights, which matter deeply in a legal environment that is increasingly hostile to abortion access. In early summer of 2019, Rhode Island passed a law effectively enshrining *Roe v. Wade*, ensuring a legally protected right to abortion.

As a fourth example: impossibly expensive daycare bills often stand in the way of people parenting as they would wish. According to the Economic Policy Institute, as of 2019 childcare in Alabama cost $5,637 annually for an infant, and $4,871 for a four-year-old.[38] As Katha Pollitt notes in the *New York Times*, that's 69 percent of the average rent in the state, and 33.7 percent less than the cost of in-state tuition at a four-year college. In West Virginia, by comparison, the $7,926 annual bill for infant care is 32 percent *more* than the cost of college. More than half of Americans live in childcare "deserts," a percentage that includes significant majorities of Latino and rural families.[39] Relative to the rest of the world, the United States spends a woefully small amount on family benefits—coming in second-to-last among wealthy nations.

The dearth of childcare limits people's ability to have the families that they want, and inflicts financial hardship on many who do become parents. It also drives lower-earning parents—often women—out of the workforce, and provides motivation for parents to suppress the wages of care workers. Movements to offer free and universal childcare and pre-kindergarten reflect a reproductive justice orientation—notably making mainstream headlines in 2019 through Elizabeth Warren's presidential campaign.[40]

A related effort seeks to help in-home childcare providers unionize, a campaign that has already succeeded in places like Minnesota and California. Minnesota in-home family childcare workers are being organized by the American Federation of State, County and Municipal Employees, or AFSCME. The California bill allows self-employed childcare workers who serve subsidized families to collectively bargain with the state about reimbursement rates. Included in this new class of organized laborers would be trained workers, unlicensed friends, family, and neighbors who care for children when their parents are in a pinch. Together, these at-home providers are responsible for more than one-third of the young children who attend subsidized daycare—and a larger share of toddlers and infants. These efforts are designed not only to support the quality of care that children receive, but also to support those doing the caregiving.

As a final example, housing equity work is also reproductive justice in action; housing is a critical piece of anyone's ability to parent in a safe and thriving community. During the peak of the COVID-19 pandemic, 2.5 million US children were homeless (roughly one in thirty), an astonishing number in such a wealthy country.[41] Housing is also a public health issue, affecting virtually every dimension of a person's life, from academic achievement to physical health. Sixty-eight percent of domestic violence survivors become homeless at some point in their lives, meaning that when a person's safety is under threat, so is their housing, and frequently the housing of children in their care.[42] The availability of housing is of crucial importance to the health of pregnant people, postpartum people, and their children. Parenting well and maintaining a healthy pregnancy is a thousand times more difficult without a home.

This list might read a bit like a policy brief, but our idea is to illustrate how all aspects of our lives and health are intertwined, and show what some of these principles might look like as system-level solutions. As we've seen, dozens of factors contribute to an environment of reproductive justice—including things that fall outside of the narrow purview of "reproduction" itself.

None of these political issues—reproductive health, housing, childcare, food—can be fully disentangled from others. If you pull a thread in one place, the fabric puckers somewhere else. When we look at the world through the lens of reproductive justice—a perspective that emphasizes the interconnectedness of social issues, and demands universal access to rights and care—the damages done through population-focus, particularly in service of curbing carbon emissions, become clear. Reproductive justice helps us better understand the problems—how they came to be, and how they affect us differently—as well as better see the solutions.

A Brief History of Women-Led Movements

CONCEIVABLE FUTURE: HOW WE SHOWED UP AND WHY

There are enough external forces that want to decide for young people what our bodies "should"—and "should not"—do. To us, the generative move was to make a *place where people could think, listen, and talk without fear of judgment, censure, or bossing.* We wanted a space to simply share, free of the pressures weighing down upon us. We built Conceivable Future out of that vision. We aim to be inclusive to all who would wish to participate, and nondeterministic about the experiences anyone might share. Only in this kind of shared space are people able to start distinguishing—*for ourselves*—between our own desires and fears, the conditioning and pressures we've experienced, and the realities that surround us all and shape our options. For us, taking a receptive, curious starting point opens up a wide range of strategies, both for ourselves and for the people we work with.

As much as we work to move beyond limiting gender categories, though, our own organizing responds perforce to our present context, and a history of harms and exclusion. This is partly because we're cis women, and living in our bodies at this point in time has taught us some important things about the ways that our experiences are generally different from men's. We picked reproduction as an entry point for our organizing because it was an issue in our hearts, and affecting us, as women, differently than our partners, who were mostly men. We were thirty when we met, and the gender-based and sex-based expectations of the world were pressing down on us, as they do most women. But sex-and-gender-based movements, as effective as they can be, are complicated. They can reinforce very specific messages about who women are and about the essential and valuable parts of our identities.

On both of our minds when we met was how the climate movement reflected an individual celebrity culture of middle-class white men, and men's experiences broadly. Much has changed in the last decade, especially since the Standing Rock encampment, the Sunrise Movement, and the advent of School Strikes, but at the time the public faces of climate activism were Al Gore, Jim Hansen, Bill McKibben, and Leonardo DiCaprio. We were confused and annoyed by the mainstream fantasy that activists are lone-wolf heroes who lead the masses by themselves, because we knew from both history and personal experience that all successful activism is collective. And much of the unheralded work is done by women and by people of color.

Shortly before we met, Meghan had witnessed her dear friend Jay O'Hara block a coal-carrier ship from entering its port in an act of climate-concerned civil disobedience. In late 2013, Jay and Ken Ward anchored their lobster boat in front of the Brayton Point Coal Plant on the Massachusetts coast, straight in the path of the oncoming *Energy Enterprise*, announcing that they wouldn't move. The Brayton Point power station was, at the time, the single largest source of carbon pollution in New England. The ensuing trial and media attention emphasized the smallness of their vessel, the heroic conviction of the two men. One photo of the tiny lobster boat, pictured against the backdrop of the unfathomably large *Enterprise*, was reprinted in magazines and newspapers nationwide, becoming an emblem of the movement for a short time. Over the course of a few months, the narrative built—the climate heroes fought Goliath, and in a final twist, the prosecutor dropped the charges. Somerset mayor Sam Sutton joined Jay and Ken at the People's Climate March in New York City, alongside half a million activists.

As an inside observer, Meghan had a clear view of both the myth and the reality of the action. The reality included a planning and support team in the lead-up, follow-up, and during the action itself—including fundraising, lawyers, and extensive planning. It also included privilege—that invisible support—that led to the action's positive perception: the individualist white-guy myth that got reinforced.

During the trial it became obvious that gender was a key factor in the action's success—and the action was extremely successful! The David-and-Goliath imagery was important, and even the language in the press was telling, conjuring visions of hardy, principled John Waynes at sea. Jay and Ken made the most of that vulnerable moment, to their credit. But when we look at other examples of direct action where the public actors are women and/ or people of color, we see very different myths emerging.

Perhaps the most famous American woman who had engaged in direct environmental action by that time was Julia "Butterfly" Hill, who spent two years occupying the crown of a two-hundred-foot-tall redwood tree. While she was clearly hardy and principled, she was often characterized as cute but kooky: of no real consequence. Her whiteness and conventionally attractive appearance brought mainstream attention to the action, but also served to discredit its seriousness. At the same time, one of the most effective climate campaigns was being led by Rosa Hilda Ramos, in Cataño, Puerto Rico, to protect marshlands and sue polluting industries. Her organization, CUCCo (Comunidades Unidas Contra la Contaminación), won major victories, including a lawsuit against the US federal government, and she was recognized with a Goldman Environmental prize in 2008. But despite this organizing success, neither she nor the organization were household names in the early 2010s.

It seemed clear that the lobster boat action would have unfolded very differently indeed had it been anyone other than two white men on board. The cultural narrative of David and Goliath breaks down if David is a woman. Rugged individualism is reserved for men, and that has always been true of the environmental movement. It also seemed clear that the individualist myth erased the reality of group effort. Meghan's experience of watching that trial, and of supporting her friend though it, posed a certain set of questions: If two women on a lobster boat would not be greeted as heroes but rather patronized or simply ignored, what does women-led climate action look like? And what can civil and political activism be other than individualistic and celebrity-focused?

This questioning is part of what led us here. What we've learned from our years of organizing together is that this bind is characteristic of feminist organizing. On the one hand, women's experiences—our responsibilities, our physicality, and the social expectations that are held of us—are often different from those of men. For example, women are more likely to take time off during disruptions of usual childcare arrangements than men, and women do, on average, fifty minutes more of housework per day in the United States, not counting childcare responsibilities.[1] Droughts disproportionately affect women worldwide, as the responsibilities of water provision fall largely upon women, affecting health, free time, and access to education.[2] These gendered experiences obviously shape people's daily lives (not to mention their long-term earning potential, free time, and life satisfaction). Women and femmes are disproportionately victimized by sexual violence. Recognizing these differences, and how they matter, is the basis for much of the feminist work that has happened over the years. Women have different experiences of parenthood, of work, of community organizing, and of the climate crisis. These differences obviously matter.

On the other hand, there is tremendous variation even *within* women's experiences, and feminism faces the persistent challenge of keeping ourselves from becoming essentialist, reductionist, or exclusionary by defining our experiences too narrowly. For instance, although the gender pay gap in the United States remains a defining factor for many women's lives, that gap varies by race: Asian women make, on average, 90 percent of what white men do; white women 79 percent, Black women 62 percent, and Hispanic women 54 percent; women do not all experience the pay gap in the same way.[3] Those of us who are cisgender are at lower risk of sexual violence than our transgender sisters. Women are not a monolithic group. Although there are identifiable similarities, there is a lot of variation in our lives.

So too with reproduction. We knew when we began Conceivable Future that any conversation about childbearing or reproduction was apt to be controversial in the highly polarized years of 2014–2015. The sensationalistic coverage that we got in the first few years confirmed for us that this was a trigger point—and that trigger points can indicate painful rifts as well as pathways to healing. Alt-right news sources called us names. We declined to go on *Tucker Carlson Tonight* (twice). We made—and held—a nonjudgmental, non-prescriptive space for people's real, varied experiences. This strategic gamble paid off in some important ways. We have helped nudge this conversation into the mainstream, and even if the conversation missed the nuance that we wanted, some of the nuance has come with time.

In time several other groups began organizing at the intersection of the climate crisis and reproduction. The first, in 2018, was Birthstrike, founded by musician and activist Blythe Pepino in the United Kingdom. Next was the No Future Pledge, organized by Canadian teenage activist Emma Lim. The advent of both groups was interesting for Conceivable Future because they were *actually doing* what people incorrectly believed *we* had been doing all along: pledging not to have children unless and until their leadership takes major climate action. They became lightning rods for a lot of the negative attention we'd been receiving, and we suddenly became the milder, more palatable form of this activism.

In her essay "Birthstrike: A Story in Arguments," Jessica Gaitán Johannesson—one of the organization's leaders—writes "by focusing on the alarm, we failed to bring attention to the kind of action needed, and by doing so at a particular moment in time, when there was a wider public awakening to the climate crisis in the Global North, we opened ourselves to co-optation."[4] Both groups expressed nuanced arguments about their actions, but because of the sensational nature of the pledge/strike strategy, and because the mainstream conversation obliterated nuance, the caricature of their concerns is what persisted. Neither group is active today, although most of those involved are still active in climate work.

When we looked at the history of women-led social movements, we found countless powerful—but essentialized—examples of sex-based and motherhood identity-based strategies. The birth strike, or no-baby pledge, has its precedent in sex strikes, which we'll examine a little further on.

But we want a world where women's political vocabulary is bigger than our sexual or parent identities, and wherein our definition of womanhood is, too. A sex strike assumes heterosexuality as the default, and reinforces the idea that men are driven only by uncontrollable hormones instead of values. It reduces women's political power to our ability to simply avoid going to bed with someone, and leaves us outside the closed doors of direct politics. It assumes that the only way we can be powerful is by *influencing* the powerful. Defining us by our sexual or reproductive capacity can harm women by objectifying, infantilizing, and depoliticizing us. It harms men by discouraging them from identifying and participating as parents, partners, and caregivers, reinforcing a limited understanding of men's identities. And it harms trans and nonbinary people by upholding a dated biological essentialism. It offers only an incomplete understanding of what our power is, and where it can be found.

That's not a world we want, either.

This bind became particularly clear to us when the press coverage of Conceivable Future started taking off. We saw dozens of variants of the same title, something like "Meet the Women who're Choosing not to Start Families Because of Climate Change." This was the repeated misunderstanding, that we were advocating an anti-reproductive strike or pledge. We want to unpack this, because it's dense, and it points us at some of the big issues that emerge around gendered organizing.

First, we're not all women. The spokespeople for our group are two white cis women, but we work with people of all genders, and the climate crisis is clearly shaping the reproductive and family decisions of people all along the gender spectrum. Second, Conceivable Future represents parents, parents-to-be, the undecided, people who are child-free by intention or by circumstance, grandparents, non-grandparents, and members of chosen-families. A family can look infinite ways and, to that end, two or more people are a family. Third, and as we will continue to argue throughout this book, no one makes choices freely in this crisis. The reproductive justice tradition has taught us to be skeptical of "choice" as a way to talk about the exercise of human rights that are not equally enshrined or enjoyed. Finally, we don't believe the outcome of the decision—to have children, to not have children—is relevant to the political meaning of this decision.

We've found that the Conceivable Future conversation—and the terms of it—resonate very intuitively for white, middle-class-ish women, even though the climate concerns themselves are more widely shared. For many people within this category, fearing for the future of their children represents two big departures: first is the departure from conventional family expectations that many have experienced (grow up, go to college, get married, have babies). Second, and probably more importantly, it is the first time that many people in this group have confronted, head-on, the ways in which the prevailing power structure does not value our children, even if we are used to the ways in which it does not value us, as women. And so we have needed to learn to speak from a place of inquiry and self-reflection, of acknowledging how feelings, power, and identities all entwine.

One of the limits of the strike/pledge movement is that given its particular resonance with white middle-class activists, it becomes a call for the status quo: we won't have babies until it's safe for *us*. The "us" may not be intended to exclude, but if the only people making the demand belong to a majority-white middle class group, the conditions that would return a feeling of abundant future to that group would do little to change the present-day harm against everyone else. We don't want a world that's only safe for middle-class babies; we want a world safe for *all* babies.

Another of the limits of this kind of demand has to do with the particular political invalidation that white middle-class women are subject to. This is different from being oppressed or ignored: this group is often very publicly dismissed as unserious. That is, because we are protected by our racial and economic status and location in the United States, we are often seen as having "first world problems," despite the violence and exclusion that remains present in our lives. Bad-faith commentators use the fact that other groups suffer more as a sleight of hand to obscure our valid concerns. The underlying message is "What could *you* possibly have to complain about?"

Blythe Pepino of Birthstrike suffered much more of this kind of public ridicule than we at Conceivable Future did, partly because Birthstrike's strategy involved seeking mainstream exposure, but partly because she harnessed her own, vulnerable experience directly to her political demand. Some analysts suggest that Birthstrike ended because of a tension between individual reproductive decision-making and the much broader "ask" of rapid structural change from governments and corporations to arrest the climate crisis, which observers have argued ultimately confused Birthstrike's message.[5] In other words, they thought that Birthstrike's demand was too big. This may be so. They also had the temerity to demand change as *beneficiaries* of the status quo, and were punished for it. The history of women-led social movements shows us that these consequences are depressingly predictable.

Beginning by listening

We have the capacity both to experience the individual consequences of things (e.g., climate change) as well as to understand those consequences within a history, a context, and within a political conversation. There is a big difference between starting a conversation with an *open* question and ending a conversation before it has begun by asserting a single correct answer to an *impossible* question. Weaponizing choice did not feel good to us as a strategy. Instead, we have tried to make space for the great variety of personal experience, and to use that to build some power.

More broadly, we believe that inclusive movements begin with an invitation rather than a series of rules. This is why we start with a question: "How is climate change shaping your reproductive life?" Asking this question leaves space for an array of answers, as Part I explored.

And we need to make a very important distinction here: we do not use our reproductive capacities *as political ends*. Rather, those capacities are an entry point to understanding our experiences, building relationships, and gearing up for greater action. In other words, we don't think political power can or should come from how many or few children people have. That prescriptive approach accepts the historical limits that actually need changing and erects barriers between potential allies. Instead, thinking and talking about the future and about our families within it is a way to make common cause.

Ultimately, we have tried to build a framework for understanding these intersecting issues—climate, reproduction, gender, violence, capitalism, and how all of those things connect to other pieces of identity and experience—for the long term. Building an analysis is part of what we ultimately did—but the trial, error, listening, and learning were big parts of it.

Building a shared language

When we started Conceivable Future we were also responding to an absence of feeling within the climate movement, where there needed to be a beating heart. Feelings are not the endgame, but unless we talk about them and figure out a way to live with them, they will quietly unbalance our lives. And in 2014, the mainstream climate movement was still overwhelmingly rendered in a committee-crafted carbo-babble that didn't recognize the human stakes of the crisis. There was no acknowledgment of how we humans were experiencing it, individually or collectively, and there were no tools to understand what that meant for our movement.

We've tried to create a responsible space for sharing those feelings, and we emphasize that feelings are the beginning, not the end, of the game. In *The Feminine Mystique*, Betty Friedan described getting letters from middle-class housewives across the United States in the aftermath of World War II, who all felt a persistent sense of discontent, a sense that something was amiss. Friedan called it The Problem That Has No Name.[6] Throughout second-wave feminism (which was not a beacon of inclusion, even as it moved the needle in important ways), a crucial part of the work was *simply to give the problem a name*, and in doing so, to give it political power. The problem was, and continues to be, patriarchy. But naming it created leverage, and with leverage you can move really big stuff.

At Conceivable Future, talking for years about how the climate crisis is affecting people's views and fears of their own futures has helped give *this* problem a name. Given how complex and massive our problem is, naming it takes more words, but put simply, the climate crisis is a reproductive crisis. Sharing our fears, our hopes, our grief, and our excitement is the first step toward collective liberation. We have built a critique, and with that critique we aim to change institutions.

In other words, feelings are an entry point to understanding what we have on the line. In the time that we've been working, climate grief has moved from the wings to the main stage, and in the process it's been politically whitewashed. In 2014 there was no conversation about it; by now, it's in danger of being medicalized and reified into neoliberal meaninglessness. We dedicate the first chapter of Part II to understanding climate feelings. But for now, we will simply point out that acknowledging feelings can be either a cul-de-sac—a dead end unto itself—or an on-ramp to the larger structural issues related to climate change: where the problems come from, who loses, and who benefits.

A lot of our first steps were gut-checks and community-checks about how our analysis landed with different participants. We tried our ideas out on the eternally patient people in our lives to see if they floated. We tweaked, shifted, rephrased, reoriented. (That, by the way, is also an important organizing lesson: take the feedback.) We read a lot, and went down a lot of internet rabbit holes. We stepped in some puddles, and avoided some puddles. We pushed ourselves to be brave and speak our truths, while listening to the truths of others.

We hope that this book can do what our organization also strives to do: offer a framework for understanding, a place to put feelings, and starting points for getting involved: a flexible form.

HOPE, SEX, AND POWER: A VERY BRIEF HISTORY

We've always described Conceivable Future as a *woman-led* network. We do this simply because we are two women who run this organization, whose organizing responds to a history of gender-based (and gendered) movements worldwide. We're not the first to enter this space and we likely won't be the last; we're here at a certain time and place, and using tactics that make sense to us.

As women who organize discussions about reproduction, we have necessarily joined a legacy of women's movements. We use that term broadly to mean movements that are led by women, or are designed to improve women's lives, regardless of who else is participating. We also recognize that even the word "woman" evokes different resonances at different times and places in history. Because we are part of this tradition, it's important for us to consider how our work resonates within it, as well as the ways in which it diverges.

Framing women's political movements as deriving from our sexual, gender, or reproductive selves has tremendous power and also serious limitations. It can amplify voices that have historically gone unheard, bringing the lives and struggles of oppressed groups into the mainstream. It always intersects with other parts of people's identity and experience, whether that is race, religion, nationality, or class. It can have shock value, jolting private or taboo experiences into public discourse and forcing societies to do some self-reflection. And it draws from a reserve of largely unconscious cultural symbolism to enable new, and sometimes deeper, conversations about what we value and why. But whether this framing is intentional or unintentional, pro or con, it edits our identities down to our reproductive and/or sexual capacities, undermining our self-sovereignty and making it harder to exercise the full range of human rights that belong to us. It also perpetuates the false beliefs that gender is biologically determined and that reproduction pertains only to women, so is not men's concern (and therefore not of universal significance).

In short, it's a power move—but a complicated one.

Sex as Power

Women have used their sexuality as a political tool throughout the history of gender-based oppression. The Greek play *Lysistrata*, originally performed in 411 BCE, tells the story of a woman who organizes other women to withhold sex, ultimately forcing an end to the Peloponnesian War. The women of the Iroquois nation withheld sex in the 1600s to ensure that they had veto power when deciding whether the tribe went to war. Those women won their seat at the table; their effort is one of the first known women's rebellions in the United States. Liberian women catalyzed the end to a fourteen-year civil war that killed over 250,000 people by staging a sex strike in 2002. In April 2009, ten Kenyan women's groups put out a call for women to abstain from sex in order to end disputes between the president and prime minister. The sex strike lasted a full week, ending with a joint prayer session when President Kibaki and Prime Minister Odinga finally agreed to talk. In 1997, Colombia's military chief called for the wives of paramilitaries, guerrillas, and drug lords to withhold sex in a bid for peace. And in 2006, wives and girlfriends of gang members in the town of Pereira reportedly withheld sex to force a truce.

It is not so much the outcomes as the tactic that interests us here.

A sex strike acknowledges the reality that, for much of their lives, many women's primary experience in society is of sexual objectification. It is meaningful that all of these instances involve women demanding an end to violence using their bodies: sex strikes have typically been used to compel a cease-fire or to exercise a tempering power in cases of armed conflict. Leaders of these movements all capitalize on the roles traditionally ascribed to women (usually in patriarchal societies) to enact their anti-war stances.

And these strategies have worked precisely *because* of those roles. War has traditionally been gendered: men are seen as brave warriors and women are simultaneously the cause, the prize, and the victims of war, either the embodiment of home and nation, the spoils, or the mourning widows. But if war is gendered, so is peace: the identities of "woman" and "mother"—life-giving roles—are especially important in times of war, because they offer a contrast to death and destruction. Women's symbolic roles as moral guardians and keepers of life can be deployed to legitimize their political demands.

The sensationalism of a sex strike garners media attention, which in turn helps propel the cause into the spotlight, but whether these strikes have been wholly "effective" is still the subject of some debate. Even Leymah Roberta Gbowee, one of the leading Liberian peace activists, has publicly questioned the efficacy of the sex strike compared with other direct action (such as mass demonstrations and sit-ins).

Although sex strikes are a form of collective action, they are also an indirect form of power: women use their capacity to offer and withhold sex *as a way to pressure men to do something*. Just as a sex strike to demand peace necessarily occurs within conflict, it also necessarily occurs within an unequal power structure in which women do *not have a seat at the table*. Inequality, exclusion, and reductive essentialism mark the political reality for many women, and a sex strike is an expedient, potent tool under those circumstances. But it can also reinforce the inequality, and experience of objectification, from which it emerges.

In other words, organizing around women's sexuality is strategic, given the power structure under which many women live. At the same time, it can be essentializing and limiting.

Sexual Violence

Movements that withhold sex—or babies—for political change do so in a world where sexual violence is ubiquitous as a tool of subjugation. That is: these tactics arise from women's experiences of the deep and inherent link between sexual violence, capitalism, racism, and misogyny. This stuff is old and it's in there deep. Sex-based political actions aim to reverse a narrative, to wield that age-old power of sexual oppression against the patriarchy.

Nazi Germany provides extreme examples of nearly every way both sex and reproduction can be weaponized. Genocide, infanticide, and sterilization were mobilized to destroy enemy populations, and everything from cash incentives and prizes for multiply fertile mothers to "baby farms" for unwed pregnant blonde women were used to increase the favored population. As we saw in the last chapter, the United States is familiar with many of those tactics, and the reproductive justice movement is the most organized, most informed pushback against them.

Rape, and the threat of rape, is on every page of world history. The word "rape" derives from the Latin *rapere*, meaning "to steal, seize, or carry away." Seizure and rape were accepted methods for men to "claim" a wife in early history (and remain so contemporarily, in some places).[7] If owning property and gaining wealth are considered marks of manhood then, according to journalist Susan Brownmiller, "Concepts of hierarchy, slavery and private property flowed from, and could only be predicated upon, the initial subjugation of women."[8]

Rape as warfare

Mass rape is most common as a tool of warfare; here, men use rape as a political weapon. It becomes a *political* weapon when it intends not simply to harm and dominate individuals, but to intimidate and damage entire communities of people. Rape campaigns capitalize on social and gender stigmas to demoralize and destabilize whole communities, weaken social ties, and invalidate the rights and autonomy of its victims. Some militaries have used rape explicitly to impregnate their enemies, or to damage their reproductive health. When men use rape, the weapon continues to injure people beyond the primary victim: the target is society and bodies are the battleground. It is not accidental that the most common target bodies belong to women. These forms of violence enforce the view that women are political and demographic commodities.

There are many illustrations of rape as a weapon of war across arts and literature, which represent—more than anything—just how taken for granted it is. Examples include: Homer's *Iliad* and Giambologna's sculpture *The Rape of the Sabine Women* (1574–1582) in Florence's Loggia dei Lanzi. Rape is noted throughout the Old Testament of the Christian and Hebrew Bible, including: "Women are raped in Zion; virgins in the towns of Judah" (Lamentations 5:11), and "For I [God] will gather all the nations against Jerusalem to battle, and the city shall be taken and the houses looted and the women raped" (Zechariah 14:2). One of the greatest historical instances of mass rape warfare dates back to the thirteenth century and Genghis Khan, who amassed an empire across Asia and Central Europe. Khan established his power, in part, by creating specific policies of rape warfare against millions of women and young girls. He told his courtiers: "The greatest pleasure in life is to defeat your enemies, to chase them before you, to rob them of their wealth, to see those dear to them bathed in tears, to ride their horses, and to ravage their wives and daughters."[9] This sentiment is not lost to history: it was paraphrased in the 1982 Arnold Schwarzenegger film *Conan the Barbarian*.

Rape was also used strategically in World War II by the Soviets and Japanese (as was the case with "Comfort Women" and the infamous Rape of Nanking), as well as the Nazis. Rape of Vietnamese women by US troops during the US invasion of Vietnam was a "widespread," and an "everyday occurrence" that was "condoned" and encouraged by the military, and had its foundation in military training and US culture.[10] The last decade has seen a growing number of civil conflicts around the world increasingly target women and girls, as well as people who are seen as transgressing gender norms, leaving the number of rapes and forms of abuse at alarming levels.

We can confidently assume that anywhere there is war there is rape, and the whole truth of sexual violence as warfare can never be known. But a list of countries where rape has been *documented* as a weapon of war in past and present conflicts indicates its scale: Afghanistan, Algeria, Argentina, Bangladesh, Brazil, Burma, Cambodia, Canada, the Central African Republic, Chechnya, Congo, Cyprus, Democratic Republic of Congo, East Timor, El Salvador, Guatemala, Haiti, India, Indonesia, Kuwait, Liberia, Mozambique, Nicaragua, Peru, Pakistan, Rwanda, The Sudan, Sierra Leone, Somalia, Turkey, Uganda, United States of America, Vietnam, the former Yugoslavia (Bosnia, Croatia, Kosovo, and Serbia), and Zimbabwe.[11]

In order to imagine a future of reproductive and sexual autonomy, we need to be clear about how the past underwrites the present. The power of sexual violence propels virtually all of history, while, because rape is so underreported or documented, the number of victims will remain unknown forever.

Being under-told will never make these stories less important.

Sexual violence in civil society

Men also use sexual violence and rape in a less organized way: typically to shame, silence, or literally beat down women who are seen to challenge norms. This, as well, is an old story: rapists have used sexual violence to punish women who were suspected of infidelity; white men systematically raped and sexually assaulted enslaved Black women in the United States to subjugate them and reinforce racial hierarchy.[12] In some cases, today, incidence of rape increases alongside women's educational levels and women's job prestige; trans women continue to experience by far the highest levels of violence in the United States and elsewhere.[13] Threats of gendered violence are also the norm online. Twitter went violently berserk when writer Jessica Valenti *simply asked* whether there was a country in the world that offered menstrual products for free; she was overwhelmed with humiliating tweets by men insulting her body and promising sexual violence. Several years later when menstruation activist Cass Clemmer made an online post about their experience as a genderqueer menstruator, they too got thousands of threats of death and violence.

In other words, there's a backlash: when people—especially women and femmes—threaten power structures by attaining too much, by breaking out of prescribed social roles, or by making demands, we are often punished with sexual violence. And this means that we can't talk seriously about gendered movements unless we also acknowledge the gendered anti-movements, and the violent dangers that can await people who challenge the status quo.

Mother(s') nature

If sexuality has historically been one powerful part of women's political identities—and if sexual violence has been a way that men can exert their own power, or take ours from us—then motherhood is another piece of our cultural identity with proven political power. As we discussed in chapter 1, the climate crisis poses a full range of threats to the lives and health of children. And although society is talking more about care work and its cultural value, the expectations for that work still fall largely on women: the "valuation" of care work has not yet shifted how gendered power structures work. Earlier we named some of the groups that are already organizing for climate action within the motherhood frame: Mothers Out Front, Moms Clean Air Force, ClimateMama, and others. Like the movements that came before them, these groups leverage the symbolism of patient strength, selflessness, and ferocious protectiveness that are inherent in our ideas of "motherhood." These groups are the organizational embodiment of the fabled mother whose overpowering love allows her to lift a car off her baby. This ideal continues to carry cultural power, even in the cynical political atmosphere of the twenty-first century.[14]

None of this focus on love and tenderness, however, is to say that mother-led movements don't go into the streets; they are often overtly political and confrontational. Wielding motherhood in public protest can add emotional depth and mainstream legitimacy to political demands and direct action. It also escalates the natural drama of nonviolence: it's hard for a society (or a police force) to look moral and legitimate while arresting and battering mothers. Recent history is full of examples of women mobilizing the political and cultural power of motherhood to make political demands. These movements arise both from the emotional truths of parenthood and also a form of performance—an amplification of a role. In these actions, a mother's individual personhood can be subsumed in the archetype. Like sexuality, motherhood can be a powerful tool, and full of complications. The identity of "mother" carries a unique moral gravity.

Because of that, examples of this kind of organizing abound, and the symbolic heft comes in part through making the private (e.g., that which is "family") public. In El Salvador in the late 1970s, for instance, a number of women's groups organized themselves as CoMadres. They began as mothers of political prisoners who sought the same truth: the whereabouts of relatives who had gone missing or been assassinated during the Salvadoran civil war. The group embraced many means of protest: occupying the Red Cross and churches, for instance, in addition to holding public demonstrations. They persevered through the kidnapping and torture of several of their own members and families, and two bombings of their office.

A central tactic of mothers' occupations is to turn personal pain into a political demand. Women often face particular

challenges when occupying public space because men are ideo-
logically associated with the public sphere, whereas women are
seen as belonging to the home—to the private sphere. The Italian
language offers a succinct example of this double standard: *un
uomo di strada* means literally "a man in the street," as in passerby,
or citizen. *Una donna di strada* means literally "a woman in the
street," but is widely understood to mean a sex worker—the least
protected category of sexualized woman. This means that women
are often very closely scrutinized for their behavior in public
spaces (which, among other things, shapes our safety in these
places). And somehow the combination—of motherhood, womanhood, and simply the act of being out in public when popular
imagery so frequently has us safely tucked away in the kitchen—is
especially potent.

Mothers of the Plaza de Mayo are another such example;
they were a group of Argentine mothers who had lost children
and grandchildren to the "Dirty War." Their goal was initially
finding the *desaparecidos* (or "disappeared"), and then to identify
those who had committed crimes against humanity, and to try
and sentence them. In early 1977, when the dictatorship was at its
most repressive, a group of just fourteen mothers gathered for the
first time in the Plaza de Mayo (the public center of the political
capital, Buenos Aires), to demand answers about the disappearance of their children. Many of their initial meetings were actually
held walking, as the police did not allow them to congregate.[15]
The Mothers, who were identifiable by the white headscarves
they wore to symbolize their missing children's diapers, were
committed to what they called a "child-centered politics," and
were a women-only movement from the beginning.[16] In order to
engage in politics as mothers, they disavowed the other kinds of
politics that they were encountering: a male politics of death and
disappearance. They used their roles as mothers to frame their
protest, demanding the rights that they saw as inherent to their
role: creating and conserving life.[17] In 2017, some half a million

protesters raised the white headscarves of the Mothers in protest against a Supreme Court ruling that would have halved the prison sentence of a member of the dictatorship and set a precedent for lenience that the people found unacceptable.

Pacifism "In the name of motherhood"

The US holiday of Mother's Day has an anti-war origin, and draws on the life-giving imagery that appears over and over again in women's occupations. In 1870, Julia Ward Howe, a white American poet and author, offered a pacifist response to the ongoing carnage of the US Civil War by issuing her "Mother's Day Proclamation," which called on women around the world to oppose war in all its forms. The appeal is full of this tension between private and public that we've been discussing, even alluding to the logic of the sex strike while demonstrating Howe's feminist belief that women have a political responsibility to the societies they live in.

It would be decades before the United States officially began celebrating Mother's Day, and by the twenty-first century, the anti-war spirit of the original movement had gotten lost: it has become a marketing holiday that leverages family love to boost consumer spending and corporate profit. Now that Rosie the Riveter, maker of fighter planes and tanks, is the face of commodified feminism, we tend to forget that many early feminists were anti-war activists. But Howe's words resonate with feminists, pacifists, and indeed what we might call environmentalists, in equal measure:

> Arise then . . . women of this day!
> Arise, all women who have hearts!
> Whether your baptism be of water or of tears!
> Say firmly:
> "We will not have questions answered by irrelevant
> agencies,
> Our husbands will not come to us, reeking with carnage,

For caresses and applause.
Our sons shall not be taken from us to unlearn
All that we have been able to teach them of charity,
mercy and patience.
We, the women of one country,
Will be too tender of those of another country
To allow our sons to be trained to injure theirs."
From the voice of a devastated Earth a voice goes
up with
Our own. It says: "Disarm! Disarm!
The sword of murder is not the balance of justice."
Blood does not wipe our dishonor,
Nor violence indicate possession.
As men have often forsaken the plough and the anvil
At the summons of war,
Let women now leave all that may be left of home
For a great and earnest day of counsel.
Let them meet first, as women, to bewail and
commemorate the dead.
Let them solemnly take counsel with each other as to
the means
Whereby the great human family can live in peace . . .
Each bearing after his own time the sacred impress, not
of Caesar,
But of God —
In the name of womanhood and humanity, I
earnestly ask
That a general congress of women without limit of
nationality,
May be appointed and held at someplace deemed most
convenient
And the earliest period consistent with its objects,
To promote the alliance of the different nationalities,
The amicable settlement of international questions,

The great and general interests of peace.[18]

Although Howe's original Mother's Day proclamation has been buried under a century of cards and chocolates, the connection of motherhood to peace persists in political and organizing movements around the world. Most of these groups rely on the same archetypal roles—of women as life-givers—that sex strikers do. For instance, a US anti-war movement from 1939 to 1945, the Mother's Movement included ultra-conservative, isolationist women's groups across the country. At its height it consisted of fifty to a hundred loosely confederated groups, with a total membership that may have been as high as five or six million. Members were typically white, middle class, middle aged, and Christian. Another Mother for Peace, a Vietnam-era anti-war group also used the imagery of motherhood to make pacifist demands. The Four Mothers group in Israel played off the same ideas; the name is in reference to the Biblical matriarchs Sarah, Rebecca, Leah, and Rachel. It was an Israeli protest movement which was founded in 1997 following the Israeli helicopter disaster, by four women residents of northern Israel and mothers of soldiers serving in Lebanon whose goal was to bring about an Israeli withdrawal from the security zone in Southern Lebanon. These movements define and instrumentalize motherhood as synonymous with peace.

Giving, sustaining, and defending life

In recent years, a wide range of women-led movements focus on *life*, if not explicitly on motherhood. These movements have been less deterministically based on ideals or archetypes, more guided by the full human experiences of organizers. One well-known example is the #BlackLivesMatter movement. In 2013, three Black women—Alicia Garza, Patrisse Cullors, and Opal Tometi—created the Black-centered political movement building project called Black Lives Matter, which began with a social media hashtag, #BlackLivesMatter, after the acquittal of

George Zimmerman in the murder of seventeen-year-old Trayvon Martin in 2012. The movement grew nationally in 2014 after the deaths of eighteen-year-old Michael Brown in Missouri and Eric Garner in New York. Since then it has established itself as a worldwide movement—particularly after the death of George Floyd at the hands of police in Minneapolis, Minnesota—and has led demonstrations worldwide protesting police brutality and systematic racism that overwhelmingly affects Black communities. Like movements specifically emphasizing motherhood, #BlackLivesMatter—created and led by women—emphasizes *creating the conditions for life*, in the face of the state-led violence that threatens it.

A similar group organizes around parent identity for the same reasons that CoMadres, Howe, and women's anti-war groups did. *Mothers of the Movement* is an American advocacy group of mothers whose Black children were killed by police. Moms Demand Action for Gun Sense in America organizes for gun control on behalf of children threatened or killed by gun violence. Mothers Against Drunk Driving focuses on a different threat to their children's lives, but all three groups find moral power in our shared cultural understanding of motherhood, and they use it to demand an end to harm directed at children.

REPRODUCTIVE SOVEREIGNTY AS A FOUNDATION

All of this leads us to one very simple insight: that reproductive, bodily, and sexual sovereignty is the foundation of justice work. Very obviously, the right to determine the number and spacing of one's children, to terminate pregnancies, and to control who has access to our bodies and when, has always been a global feminist struggle. That male domination is a fact of life for most people who aren't men means that the first contested territory for power is our own damn bodies. This autonomy is distinct from motherhood and sexuality in that the central identity is *woman as human*, entitled to full rights. Turning "mother" and "sex object" into political identities—or political tools—can elide important differences in experience along a lot of lines (race, class, cis/trans, and cultural lines).

Nevertheless it's a political struggle whose arena is pregnable and nonconforming bodies, and whose prize is self-determination. It's worth observing here that unless we're talking about fortresses, the word impregnable can have the slippery second meaning of "susceptible to impregnation, as an egg."

Any future is rooted in the soil of history, but we can't let ourselves be fully determined or bound by that history. We're at a turning point—a moment of radical possibility—in which we can imagine a future that repairs the harms of the past. That future, in which we each enjoy self-sovereignty and mutual accountability, is built from the polyphony of our experiences, rather than by roles we are forced to play. We can trade the myth of individualism for the beauty of intersubjectivity. The threat of violence doesn't determine our identities or our strategies, although it informs them. The future is unwritten, and so are our roles in it. With mutual respect, self-determination, and accountability, we make the space for our full selves.

PART II

ROLLING UP OUR SLEEVES

THIS IS THE PART OF THE BOOK IN WHICH WE GET TO WORK. In Part I we shone light on the reproductive aspects of the climate crisis; some relevant history, and the moral, political, and medical problems that arise with rising temperatures. Now we're going to look at what we can do about it. Our action strategy has to confront the unique challenges posed by this crisis. We begin with the cognitive challenge of its inhuman scale. The climate crisis is an aggregation of injustices and harms over centuries, whose growing mass compounds the injustices that caused it. No wonder our minds recoil.

Philosopher Timothy Morton named this phenomenon—of a concept too massive and complex to fully conceive—the *hyperobject*. Even those of us who arrive at the base camp, ready and willing to climb, struggle daily with this scale problem. We are so small. Time is so short. The harms are so many, so varied, taking place *everywhere*. We can become frozen in fear. Despair takes hold, or anxiety. When we reach the limit of those emotions, we can become jaded and cynical. If we don't see anyone sharing our concerns, and especially if we spend our days pretending things are normal, we become alienated and isolated.

Because very little mainstream attention has been paid to the activism people are already doing around the world, it's been easy to miss. The prevailing narrative has long been that "people just don't care," or "we're not doing anything about it," and more recently, "what's the point if we're already doomed?" In fact, a robust majority of Americans are worried about the climate crisis, but after decades of being offered false nonsolutions we are struggling with the question "what can I *do*?"[1] And because in our lifetimes we've witnessed the dismantling of both the social and intellectual infrastructure of collective action, we're asking "what can *I* do," not "what can *we* do?" Experience—and data—tell us that hope lives in action. And that effective action is rooted in loving relationships. So from love-based action, and action-based hope, we map an organizing strategy.

THE BIG NO AND THE BIG YES

Our theory of change begins with personal, internal work, and moves to family and community building, through which we create the biggest solutions we can imagine. By taking these steps, we can both begin to understand the crisis on its own terms and stay rooted in the human scale, where our lives retain their full meaning. We call it a Theory of Balanced Climate Efforts, and you can see it illustrated in this pie chart.

The mix in this pie chart is something we've developed over the course of many long conversations: understanding that in order to get to the world that we want, we help good things happen, stop bad things from happening, attend to the relationships that sustain it all, and care for our own joy and well-being.

We are framing this loosely as "the Big No" and "the Big Yes," although, like any good binary, these categories dissolve into each other. All the great work has elements of both. "Yes" and "No" may describe your orientation as a person as much as any particular action you're inclined to take. A "yes" person might orient toward good things we need more of, toward consensus and compromise. A "no" person might find more purpose in stopping bad things from happening, toward challenge and conflict. One is not better,

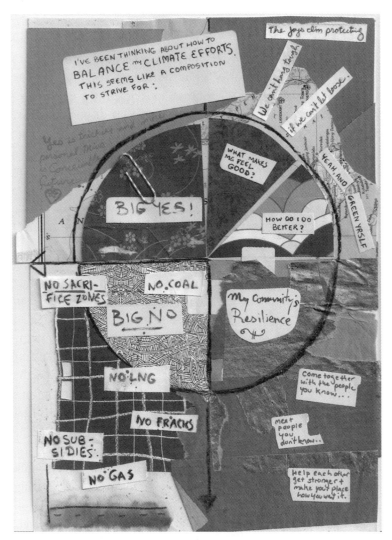

Figure 4.1

or more "effective" than the other; each orientation has pitfalls in its untempered form. People can reverse polarities countless times throughout their lives. We can also feel very "yes" in one aspect of life and quite "no" in another.

Importantly though, a "yes" kind of person is not necessarily an optimist, nor is a "no" person a pessimist. It's simply about where your proclivities lie. If optimism and pessimism have to do with our beliefs about a future outcome, "yes" and "no" are about our dispositions; what we tend to feel, and the experiences we actively seek out, in the here and now. Our dispositions may prefigure our perception of the crisis and our politics. When we learn to dance with our dispositions, we find ways to stay engaged that are emotionally tolerable, to find the comrades, information sources, and projects that flow with, not against the grain of who we are. This may sound a bit like a horoscope, but over time we've found that people often do have a good intuition of their own "yes" or "no" -ness.

Knowing your disposition doesn't always mean that making a balanced effort is a straightforward practice, though. Conflict is intrinsic to the climate movement—perhaps even more than other social movements—because so many people are trying to do such different kinds of things all at the same time, for many different (but ideally complementary) reasons. For us the Yes/No distinction is helpful because it offers a place to start without being rigid or dogmatic. It also works at any scale, from our moods to the essential truth of the change we need, which is that we have to do both all the damn time.

One problem with having idled so long in the denial discourse is that the United States mainstream didn't have to talk about solutions to the climate crisis—it was too busy debating whether or not climate change was "real." And just because we all agree that climate change is real, "people" caused it, and it feels bad, does not mean we agree on what we need to do to fix it. The Big No/Yes are about creating a general collective understanding of what we can't live with, and what we need in order to live. We don't have to agree about everything to be allies, comrades, accomplices, or colleagues, but we do need to arrive at some common values and goals.

The "Big" in our Big Yes and No also helps us overcome some of the personal obstacles that our dispositions present. If I tend toward rigidity and hard-line positions, my commitment to my group and our vision, may help me soften enough to compromise when it's needed. And if I *really* avoid conflict as much as I can in my everyday life, the support of my group, and the beauty of our vision, may give me the fortitude I need to interrupt, object, or butt heads when it's important.

Although we'll progress from the individual to the collective to Yes and No solutions in the following chapters, these different efforts are really simultaneous, and mutually reinforcing. No one can be in community, or alone—or with their family!—all the time. Nobody can survive on a diet of negatives. Nor can we sustain a permanent state of false positivity by ignoring real risks and harms. Here we return to Solnit's ideas of hope: the future is undetermined, and although we know we are in for difficult times, we still have agency.

But how does our movement become large enough, and powerful enough, to change our course in the time we have? Erica Chenoweth, a political scientist at Harvard University, has shown that nonviolent civil disobedience is not only the moral choice; it is also, by far, the most effective way of shaping politics. Looking at hundreds of campaigns across the globe over the last century, Chenoweth found that nonviolent direct action campaigns are twice as likely to achieve their goals as violent campaigns. In case study after study, in any political context, the magic number was 3.5 percent. It only takes around 3.5 percent of the population actively participating in protest to ensure serious political change, even regime change.[2] (For a little context, 3.5 percent of Illinois is about 316,000 people.) This is a heavy but doable lift. This is a meetable challenge.

We believe it to be possible, and we think the path looks like this: be both rude and loving. Talk to everyone, get together. We shout what we know, listen, learn more, and shout again.

Part II

Turning out into the streets for public protest is only part of what we do, but it is an important part! The point of public protest is to build people power, but also to create opportunities to feel larger-than-ordinary feelings. It helps us conquer the climate-scale problem by reminding us of our collective strength, and by letting us share the big emotions. It's frightening, but exhilarating, to know that we will never have better odds than we do right now.

CHAPTER 4

The Freedom to Feel

WHEN THE BELLS RING

Politics is what you do, not what you feel. Unlike politics, feelings just are. They simply exist. At times they are drivers of our behavior, or tethers to other people. Sometimes feelings inspire us, or leave us deadened. But they are not, themselves, politics.

There has been a lot of focus on climate grief lately, and many people feel a lot of bad feelings that make activism—and just being alive—harder. Climate grief is real. Loss is real. We recognize and honor both of those experiences. But here is also where the climate movement seems to have gotten a bit mired: having discovered that climate feelings are a thing, and having gotten to the depth of existential feelings, it bottoms out.

But we believe that's where the work *begins.* We are not fixing feelings, *we are fixing problems.* The goal is not to mitigate climate grief; it is to mitigate the crisis that is causing the grief. We are individuals with feelings and those feelings can move us, but the collective is the unit that takes action, and action is what defines a political movement.

Having said that, we have found we are not able to act freely unless we first have the freedom to feel. Some unexpected moments stand out in our emotional lives as organizers. Josephine describes one of hers:

> Meghan and I were standing on my mother's porch after an interview on a rare day spent together, having our picture taken by the reporter and catching up in the affectionate, shorthand way we've honed over years of long-distance collaboration. It was the sort of mild August afternoon in which nothing can be ill and even long discussions of the reproductive impacts of the climate crisis won't spoil it. Greta had just set sail to the US, and the national conversation was suddenly brimming with climate concern, which was all we'd ever hoped for. I mentioned that some longtime activist friends were having paradoxical reactions to this development; frustration, fatigue, anger. Meghan said "I'm just really sick of grief, you know?" And immediately I *did* know. I'd been dragging a garment of heavy, old emotions behind me for so long, and on that sunny morning, at the sound of those words, I became ready to drop it.

In *The Varieties of Religious Experience*, William James describes this phenomenon with peerless clarity:

> We have a thought, or we perform an act repeatedly, but on a certain day the real meaning of the thought peals through us for the first time, or the act has suddenly turned into a moral impossibility. All we know is that there are dead feelings, dead ideas, and cold beliefs, and there are hot and live ones; and when one grows hot and alive within us, everything has to re-crystallize around it.[1]

Many of us can point to a certain event, conversation, or story that radicalized us, or shifted our perspectives abruptly. An experience like this may be exactly why you find our book in your hands. Integrating these realizations is an ongoing, workmanlike process, but we may remember it as having appeared in a moment, like the peal of a bell, the dropping of a burden, the phase-shift from solid to molten. Our surroundings appear different to us in this new phase, our relationships shift, we have new questions, new emotions, new problems.

That bell had rung for Josephine before. The first time was reading Bill McKibben's article "Global Warming's Terrifying New Math" on a flight to Las Vegas in 2011. McKibben's argument made her recognize the climate crisis as a comprehensible problem, instead of a mostly subconscious dread. She remembers reading that issue of *Rolling Stone*, flying through a cloud of Californian forest fire smoke obscuring the Grand Canyon, as the moment she became ready for climate activism. But three years later, when she and Meghan met, she says,

> My father had been dead for one month and grief was making it impossible for me to do the activist stuff I'd been working on. Pretending a can-do attitude felt like pressing my thumb into a bruise. My old work was an impossibility. But here was a person who, like me, was looking and couldn't find the heart in the climate movement. We were both giving our passion and time to this cause because of the intense feelings the crisis provoked in us: fear, anxiety, sadness, and a breathless sort of hope that prevented rest.

We were each struggling privately to understand and express those feelings, because even in activist groups we each experienced an enforced flatness around us, a culture of dissociation. Losing a parent had made that environment totally unendurable for Josephine, but it also showed her a hidden thread that ran through her efforts: "Will I become someone's parent, and if I don't, will my family die with me? I was desperate for something I could do, and that first conversation with Meghan pointed the way. Everything re-crystallized around it."

The animating purpose of Conceivable Future was to help the climate movement speak from the heart and, in so doing, reach for bigger levers of political and social change. By 2019, when Greta sailed, the movement's heart was exposed and bleeding. Emotions were now the focus of much organizing, and much, much discussion. A selection of article titles from that moment:

- "Climate anxiety doesn't have to ruin your life. Here's how to manage it."
- "'Climate Despair' Is Making People Give Up on Life"
- "Manage Climate Change Anxiety With These Cold-War Coping Techniques"
- "When It Comes to Climate Change, Grief Is More Useful Than Empty Nostalgia"

What comes through those headlines, louder for its being unspoken, is panic. Activists had been waiting for the moment, or organizing for the moment, in which we could publicly acknowledge the emotional burden of this crisis. Now we could finally put words to the feelings, so began a frenzy to turn feelings into fuel. Could sparks of anxiety light a fire under people's asses? Could that cold moon of grief pull a new tide of people into the movement?

But feelings are funny, shifty, and as we all know, paradoxical. When we agreed that we were sick of grief at that moment, we weren't swearing we'd never feel it again. We were acknowledging that emotions change in an instant. "Turn [grief/fear/anger/etc.] into action" is a call for alchemy, not activism. You can't just harness an emotion, feel it forever, and ride it to low-carbon utopia. Feelings don't make reliable fuel for action, because they vanish, reverse, and arrest us with new priorities. Feelings are intrinsic to this work, but as we learned when feelings finally joined the public conversation, they can't be instrumentalized, they can't be selected, and they don't obey us.

Action and feeling have an intimate relationship, but only what we *do* is political. Feelings *can* propel action, but they resist exploitation. We simply need to know them; observe what we feel and care for our emotional selves, because we are—and will always be—people with feelings. This work is what we think of as "finding our core." At our core we create space for the feelings that reside there and we practice being aware of them. When we recognize and understand what we're feeling we're less likely to be propelled in weird, unconscious directions.

We've often heard from highly informed, highly motivated participants, "We don't have time to talk about feelings! What are we going to *do* about it?" Impatience is another familiar feeling; we all know that the time to act has long been upon us. But in the decades we've lived through, in the climate of silence that has been the norm, strange structures of feeling have grown to obscure our path forward. These emotional snares don't come from nowhere; they are rooted in history, fed by the same rapacious injustice that drives the crisis. And we need to observe them to understand them.

However immediately the crisis demands solutions, we confront climate change primarily as emotional beings. In other words: we experience it emotionally, whether we want to or not. That is our humanity.

If the animating question for this chapter were "What does my climate-engaged life look like?" Our answer would *never* be "you should feel _____." We've learned that getting engaged, and staying engaged, is based on regularly inquiring "how am I feeling?" We reject the belief that there are more and less "correct" feelings: instead, we observe the emotional lives we are *actually* living, and work with those. In later chapters we focus on building relationships and solving the big problems. But here we're inspecting the inner chamber of an engaged life, the core of ourselves that we move from, where we return when we need to heal, metabolize, rest, and reflect. Knowing how we feel is the deceptively simple—but not easy—first step.

We aim to describe experiences we've had, and that you may recognize, with some of the big, difficult feelings like fear, grief, rage, despair, as well as the less discussed but equally powerful boredom, resentment, and frustration. We also aim to prune some overgrown obstacles, from our focus on individual consumer "solutions" and body-size hang-ups to our apocalyptic fantasies, to the subtle forms of denial in which we all engage, some of which are actually crucial for our well-being. And we share strategies to care for our emotional well-being; we understand these practices not as an add-on to our busy, ordinary lives, but as tools we need to attend the death of our old world and assist the dangerous birth of our new one.

LANGUAGE PROBLEMS

When we begin to talk about feelings, the first obstacle is the words. Most of the language we use to describe the situation we're in is a deadly combination of scary and boring. It's like overhearing a meeting of technocrats aboard a sinking submarine. And like any words and phrases repeated often enough, these have lost the way to their meaning. Here's a list of words that sound worn out to us:

Climate Change, Climate Crisis, Climate Impact, Climate Consequences, Concerns, Looming, Threats, Urgent, Terrifying, Impacts, Unprecedented, Unthinkable, Threatened, Jeopardized, Systemic, Massive Scale, Messaging, Fighting/Grappling/Tackling/Combating Climate Change, Solutions, Solutions-Based, Solutionary, Action-Based, Climate Action.

We certainly need these concepts, and need to name them, but if you hear them or read them and feel your brain turn off, you may be experiencing the deadening power of cliché. The fact that our climate vocabulary is so hackneyed has two probable causes.

First, we just lived through an era of taboo, several stunted decades of "politeness" in which a gulf opened up between what we knew to be true and what was permissible to say out loud.

Vocabularies multiply and specify through use. They also shrink and atrophy—we have to use language so that it grows to meet our needs. If the words feel stiff and weak now, we practice describing what we're experiencing, and we listen to others describe what they're experiencing. As we speak it more we'll name it better, and more clearly.

The second probable cause for our small, shrill vocabulary is that we're talking about something that's bigger than even our peripheral vision. Timothy Morton's term "hyperobject" is in surprisingly wide circulation for a word born in academic philosophy, which tells us about a widely unmet language need. Morton defines hyperobjects as:

> things that are massively distributed in time and space relative to humans. A hyperobject could be a black hole. A hyperobject could be the Lago Agrio oil field in Equador, or the Florida Everglades. A hyperobject could be the biosphere, or the Solar System. A hyperobject could be the sum total of all the nuclear materials on earth, or just the plutonium, or the uranium. A hyperobject could be the very long-lasting product of direct human manufacture, such as Styrofoam or plastic bags, or the sum of all the whirring machinery of capitalism. Hyperobjects, then, are "hyper" in relation to some other entity, whether they are directly manufactured by humans or not.

And although Morton is clear that their term can refer to many immense concepts, in climate circles the word "hyperobject" is nearly synonymous with our crisis, whose immensity overflows its handful of terrible names.

The tiny vocabulary and the vast subject combine to make this crisis very frustrating to talk about.

When we use the phrase "climate change" to evoke a more specific point in conversation, it's like two people trying to scope each other through binoculars across a forest. We don't share the same frame of reference, we're not necessarily focusing on the same things, and there is *so much else all around us*. It can feel like we need to build ideas from scratch, sometimes even zooming all the way out to an embarrassed redundancy: "I believe in climate change."

So as we begin to talk more explicitly about our feelings and experiences, how do we work with this limited vocabulary? Many writers have used their own terms to encapsulate, or at least indicate, the crisis. When philosopher Joanna Macy says *The Great Turning* she refers to both the period of harm and the potential for transformation. Author Amitav Ghosh uses the title *The Great Derangement* to characterize the mad nature of the problem, as well as the madness that created it. Daniel Sherrell refers to *the Problem* throughout his activist memoir *Warmth* as a way to evade jargon and stay intimate with his subject. Anthropologist of civilization collapse Chris Begley, whom we interview later in this chapter, named his book *The Next Apocalypse*, turning on its head the idea that the "end of the world" is a literal end. And climate reporter Robinson Meyer cuts to the chase by addressing *the worst problem in the world*. These are a few tools made by people who needed to express new ways of thinking and feeling.

Being specific also helps us manage the scale problem: we're not always talking about the whole root directory of climate change. Sometimes we mean specifically temperature rise or carbon pollution. Or we're pointing to extractivism, or industrialization, or the kind of rabid capitalism that sacrifices whole portions of humanity for short-term gain. The more we learn about the crisis, the more we master specialized terms from science, philosophy, economics, engineering, big business. But we have to seek our own poetry. We find the writers, the songs, the friends who speak in terms that resonate with us personally. We find, remember, and use the words that keep us connected to our specific feelings, the specific meanings we intend.

And while we're talking about language, it's worth mentioning here that people use the word "hope" to mean different things, characterizing it as anything from essential spiritual bravery to political sloganeering to naive or deluded optimism. How your gut feels about the word hope is one indication of whether you are a "yes" or "no" kind of person. More on this later.

As we said in the introduction, we define hope as the belief that what we do has meaning. It matters not because of a specific, conditional outcome, but because what we do makes who we are. Sometimes we feel inspired or excited about possibilities in the future, and we certainly wish for good things to happen, but for us, and for many activists, hope is not an emotion. Rather, it ist a practice. It's a position we decide to occupy daily, even as we experience the whole human range of emotions. By this definition, hope may be what caused you to pick up this book, even if you weren't feeling sunny about future prospects.

NOTES ON SCALE

Given that the climate crisis is so big it's birthing a new language, we'd like to point out an Alice in Wonderland paradox of scale. Somewhere between the cosmos and the microcosmos live humans, on earth, in the midst of an ecological crisis. We try at home to shrink the impact of our consumption, and we simultaneously believe that nothing we might do is big enough to make "a difference." We are simultaneously too big and too small.

We've learned that our unintentional impacts (eating meat, driving to work) contribute hugely to the harm, while our intentional efforts (going vegan, riding a bike) make no perceptible difference. And that's a paralyzing place to be. This slip-sliding between too big and too small, with no stable sense of proportion, can feel terrible. We're the worst at every size, feeling guiltily immense for contributing to such a vast problem, feeling hopelessly small when trying to fix it.

These are misperceptions, not facts. Still, it takes practice to sustain a human-size perspective, and to return to it when we accidentally zoom too far out or in.

As individuals, our impacts, and our efforts are proportional to each other. There's only so much good or bad we can do by ourselves. Each of us is born into a society that has its own energy sources and consumption norms. It takes collective action to change these sources and norms—we'll talk about that in a moment, but first we want to point out that what we do alone at home is always going to have the same modest, personal significance. And while we've been encouraged to focus on doing *less bad*, we can equally consider doing *more good*.

We are human size. Greta Thunberg is human-size. Jeff Bezos is human-size. Their outsize effects and impacts are really reflections of the multitude of people engaging with their ideas and/or money. In our minds they embody the #skolstrijk campaign and the Amazon corporation respectively, but in reality, those entities would just be ideas without the mass efforts of human-size people to enact them. And when we consider their "individual" consumption, Thunberg has collaborated with her family, community, sailors, and Arnold Schwarzenegger to moderate her personal carbon footprint while she does her good work, while Bezos couldn't operate his yachts without a large organization of employees. If we want our positive effects to get bigger and better, we band together. So too, we need to work together to see our negative impacts get smaller.

Any climate activist will say that a big part of our work is in expressing human-size stories and campaigns. One of the first slogans prominently used by 350.org was "connect the dots." They were speaking of connecting disparate extreme weather events to the common cause of a heating climate, but we are constantly working to make other connections visible as well:

- between environmental and social issues
- between real problems and just solutions

- between communities across borders affected by the same harms
- between local infrastructure projects and global effects
- between short-term corporate profits and long-term harms
- between well-intentioned people with money and grass-roots work that needs funding

Sometimes even staying aware of these countless, far-flung dots can overwhelm us. And doing the work to connect them often means resisting norms, understanding things at a bigger-than-human scale. Our human cognition didn't evolve to cope with—or fix—global problems. But the important thing is to remember that connecting the dots is *essentially* social, inter-personal. We can't accomplish it alone, and we don't need to think about it or feel it alone, either.

THE PERSONAL IS NOT POLITICAL: RETHINKING OUR INDIVIDUALIST BURDENS

The size of this crisis confounds the truism that the "personal is the political." Yes, second-wave feminists wisely observed the ways we experience systemic harms on the individual level, and promoted the practice of getting together to describe our personal experiences in order to understand system-level problems. But Americans raised in the hyper-individualistic, consumer-focused 1990s tend to mistake our personal choices and behaviors for political solutions to systemic problems. A depoliticized psycho-logical narrative tells us that when we get to the depth of our existential feelings, that's the work done. We've been trained to think of our power coming in units no larger than "me," when at the same moment we're facing an emergency that can *only* be solved by major collective actions.

Political problems are problems of governance and systems that affect a group. Politics is leveraging power to change systems.

Political power arises when people come together to make change. But many of us still think of politics as what we individually *believe*, not what we collectively *do*.

This approach has us stuck staring at our feet, not working with our hands. And it brings with it a lot of unnecessary effort and emotional baggage. When we focus on our own self-betterment—making carbon-impeccable purchasing choices and rinsing all the plastic bags—we fail to identify the real culprits, or the effective political strategies. And we feel guilty about it. In fact, this is a deliberate strategy of the fossil-fuel companies—blaming the user rather than the producer, in an effort to keep us feeling hopeless and entrapped.[2] Not because they care how we *feel*, but because they want us to *do* nothing.

A look at mainstream climate "solutions" offers the perfect example of how successful this misinformation campaign has been: Americans were polled about the most effective things they could do to combat climate change. The top response? Recycling.[3] We vote, bring a tote, and set the thermostat low. None of these activities is bad or harmful. We do them, and we want everybody to do them. But we don't mistake them for political acts.

When we talk in terms of "doing your part," and name things like "ride your bike to work" as a solution, it focuses us down to the individual level, away from some important questions like: Is there accessible, affordable public transit too? Are there safe routes for us to ride throughout our city or town? How much of our yearly weather is bikeable? Are the buses timely? When I get home, sweaty and exhausted at 8 p.m. from my two-hour bike commute, rattled from close calls with cars and trucks, I may have already spent all the energy I have for addressing this problem. And if I don't do it? If I drive instead I'm likely to see it as a personal failing instead of identifying the systems that we can change for everyone to have workable low-carbon ways of living and getting around town.

The popular philosophical sitcom *The Good Place* imagines a numerical moral ranking system for each person on earth—enough points and you go to "the good place." A punchline comes when Eleanor and company discover that (spoiler!) literally no one has made it to heaven in 521 years. Beginning with colonialism and continuing through the industrial revolution, even the best-intentioned individual is more and more deeply embedded in a web of exploitation, and every ordinary action causes unintended harm. The reformed demon Michael reads from the Book of Dougs to elaborate: "In 1534, Douglass Wynegar of Hawkhurst, England, gave his grandmother roses for her birthday. He picked them himself, walked them over to her, she was happy . . . boom, 145 points."[4] However, in 2009, "Doug Ewing of Scaggsville, Maryland, also gave his grandmother a dozen roses, but he lost four points. Why? Because he ordered roses using a cell phone that was made in a sweatshop. The flowers were grown with toxic pesticides, picked by exploited migrant workers, delivered from thousands of miles away, which created a massive carbon footprint, and his money went to a billionaire racist CEO who sends his female employees pictures of his genitals."[5]

There is no way to "win" a rigged game. And that knowledge lands hard when we only know how to try fixing things alone. We might come to think of our individual green habits, like hanging laundry up to dry, or switching off lamps and power strips, as the taxes we pay to disengage. But once we understand that we're not massively, personally culpable—we're just implicated in a broken system—then the individual stuff falls into better perspective. We can see the big picture, and our place in it, so it gets easier to decide how to spend our effort. We understand the difference between the personal scale and political scale, so we can start the shift from avoidance, guilt, and silence toward engagement.

"MY" CARBON FOOTPRINT: CARBON GUILT AND CARBON FRAGILITY

Another drawback to the carbon footprint mentality is that it primes us to pursue a point in our "green" "living" where we can look around and say, "well it's not *my* fault." As if the end goal were not to stabilize the seasons, but to dodge the blame. The carbon offset market monetizes this mess of feelings.

Much of the mainstream conversation about emotional responses to the climate crisis has been focused on guilt. Any climate activist can testify to the weird Father Confessor role that's been foisted on us. People want to unburden themselves: "you're going to hate this: I bought a plastic bag salad-kit," "Is this bad that I just really want to see a rhino before they're gone?" NBC even launched a strange platform called "Climate Confessions" in September of 2019, inviting participants to share "where you fall short in preventing climate change." The listed categories were Plastics, Meat, Energy, Transportation, Paper, and Food Waste. In addition to a robust troll presence, this prompt attracted responses such as: "I traded my tiny Spark for a Trax which uses more gas. Living in traffic-clogged So Cal I didn't feel safe in such a car," and "Working in a state hospital that isn't on EHR all documentation is on paper and the facility doesn't provide recycling."[6]

These are lone people shouldering the blame for entire defective systems. And as well as showcasing NBC's dreadful misrepresentation of what might "prevent" climate change, this project exposes the perverse structure of feeling we've constructed, where our focus on these perceived individual sins not only invites us to wallow in our own moral suffering, but makes us judgmental of others and blind to the real crimes. And criminals. Nobody can endure feeling guilty for long, so we often try to relieve it by tuning it out or offloading it on others.

A fruitful comparison might be to the discourse around whiteness, white privilege, and white fragility in the United States. While racism harms people of color through the unequal workings of most major institutions, many white people imagine racism purely as rude interpersonal behavior, which they themselves most certainly do not practice. Robin DiAngelo, a white woman who writes and teaches about anti-racism, coined the term "White Fragility" to describe white people's emotional responses when confronted with their complicity in racist systems.

To make the parallel, we might consider those of us in the West, especially those of us with the financial means to travel for recreation, resettle after an extreme weather event, or enjoy luxuries such as second homes or vehicles, as having "carbon privilege." You might also view older people as having carbon privilege relative to younger people. We did not make the system of exploitation we were born into, nor did we choose when or where we were born, but we have benefited from our positions.

When we consider our safe or comfortable position in relation to someone else's, the first instinct may be to feel defensive or to quit the game; *I don't want to belong to the group that is doing harm.* We may point at the harm other people are doing and reject them. We may try to withdraw from the group. From guilt, we can't draw the real conclusion: *I want my group to stop harming.* It takes a lighter, fuller heart to try to change the group without disavowing it, whether that group is white people, the Western world, the bourgeoisie, the patriarchy, or anyone else with some amassed power and some bad ideas.

It's hard not to want to disavow all of it. Compared to the rest of human history, many Americans live like little emperors. And like emperors, we're nagged by fear that our books don't balance: that with bounty comes danger. In this part of the world, hunger is the result of injustice, not scarcity. We're more often troubled by getting rid of stuff than by acquiring it. If we could afford it, we could arrive anywhere at all in less than two days. Deciding how to live one's own ordinary American life is strange work, guided by ambivalent feelings about the course that our insatiable ancestors set. We might travel less, buy less, give up entire food groups. And addicted to this bookkeeping, it's easy to mistake self-constraint for something with broader societal benefit.

As we make the comparison between climate guilt and white guilt, we're not suggesting that any emotional reaction to the climate crisis is exclusively felt by a specific racial group. But to ignore how inequality in the United States shapes people's emotional experience of the crisis would be to buy into the depoliticized individualist narrative. Dr. Sasser, who in chapter 2 described some of her findings about women's emotional experience of climate change, explains:

> Climate-related mental and emotional health issues are, in fact, a climate justice issue. So when we think about climate justice, we think about the distribution of burdens, impacts, and benefits for different communities, and how those distributions are unequal on the basis of race and class. And these emotional and mental health impacts are also differential burdens, also shaped very heavily by race and class. And I'm saying that how people feel about whether they can literally weather these storms [if they have kids], is also very much shaped in many instances by race and class too.

People who have the time and disposable income (usually middle- and upper-middle-class people) to focus on their own footprints as a hobby (check out my new Tesla!) can sometimes lose track of the bigger picture, and look down on others who can't, or don't, do the same. This is a path that some go down perhaps because it feels like something we *can* control: it becomes the judgmental coda to a panic response. The only way we tend to think about reducing others' consumption is through "educating" them (often experienced as shaming).

But from the climate's perspective, the solar panels could be on anyone's roof, the insulation could be in anyone's walls. What if, instead, we all viewed our emissions as a collective social project? What if people with means made gifts of efficiency to others, instead of only fine-tuning their own gadgets to assuage their own guilt?

In other words, if we instinctively don't want to invest in other people's carbon reductions, *we have to ask ourselves why.* If, instead of a judgy comment or side-eye, others receive a green gift or financial assistance toward an efficiency upgrade, they may be more interested in what we have to say. Judging our neighbors' consumption habits is a luxury pastime with no real-world benefit.

LOW-CARB(ON) DIET: ORTHOREXIA AND FAT-PHOBIA IN THE CLIMATE MOVEMENT

For many of us, raised in an image-focused, sexist, and fat-phobic culture, the idea of being too much, consuming too much, and the logic of the individual carbon footprint may feel uncomfortably similar to the logic of diet culture. Recognizing these bad, private feelings, and the way they've been mobilized by climate communications, can free us from painful and divisive snares.

If you read any teen or women's magazine between 1980–2010, you know this genre: to look like a model, you should wear tight jeans when you go out to eat so you can't eat too much, drink 16 ounces of water before every meal, wear heavy shoes as a simple way to burn more calories. And on, and on. We've been primed for an adulthood of finding fault in ourselves, not in systems: Ten easy tricks to shrink your carbon footprint, unplug appliances when you're not using them, wrap gifts in newspaper not gift wrap, open the fridge as little as possible!

Carbon footprint anxiety resonates with middle-class women in a special way. We are people with some privilege and access that can be used to make big demands. But simultaneously our cohort is underrepresented in the movement. Why do some of us find self-abnegation a better fit than activism? We know anti-fatness has played a big role in enforcing sexist and racist hierarchies— we're critical of it even as we're still subject to it. But we still haven't turned that critical eye on the ways anti-fatness damages the climate movement.

Orthorexia—a term coined in 1997 by Dr. Steve Bratman—describes an unhealthy preoccupation with healthy eating. This can be eliminating entire food groups, refusing to eat nonorganic foods, and so forth. People's parameters may be based on different concerns (local, ethical, vegan, keto), many of them laudable. The problem arises when it harms a person's body or interferes with a person's meaningful life. We're expanding the use of this term to include "green" behaviors that may not have to do with food or diet, which may begin as laudable ideas but come to obstruct meaningful participation in the world.

The project—to make oneself impeccable—distorts one's own perspective about the problem we're trying to fix.

This connects to our scale problems, and feeling like "my" impact needs to be smaller, as well as my appetite, my buttocks. And it becomes particularly divisive when it turns to other people's impacts, appetites, buttocks. A medical gloss and faux concern for others' health gives license to gatekeep, making specious analogies between fatness and CO_2 consumption. This is easy rhetoric whose main real-world effects are to needlessly alienate people from the movement.

Josephine doubts she'd be a climate organizer if she hadn't been a fat little kid. She explains, "hot weather made me uncomfortable and embarrassed. Fear of global warming entered my body when it was soft, vulnerable, and young. It still lives there today. I needed to decide for myself that I belong in a movement that can be harsh and unloving to people with fat bodies." She continues:

> By unfortunate coincidence, the mainstream environmental movement was coalescing around a fit, outdoorsy, personal responsibility mythos that felt inhospitable to many people, including young, fat me. The story was that some people eat too

much junk food, drive because we're lazy, buy cheap disposable everything because we don't value what's precious. Rather than constraining fossil fuel corporations, we needed to "tighten our belts." The crisis was not political, it was characterological. Even as I became politically active as a teenager, and despite my climate anxiety, it took me a long time to find environmentalism. I didn't think I fit.

And while Captain Planet's villains, Hoggish Greedly and Sly Sludge, were fat and rich, the implicit enemies of environmentalists have been the poor, making "poor" consumer choices. During the same thirty years that the climate crisis has been coming into public consciousness, the correlation between fatness and poverty has risen from nonexistent to high.[7]

The view of thinness as an intrinsic good has pervaded the environmental imagination, and though environmentalists didn't invent anti-fat bias, they have put it to work. Just as focusing on individual consumption limits our capacity to imagine systemic solutions, so cultural bias, as well as poverty, limits individuals' ability to exercise consumer "choice." We know that organic foods and leisure workouts are primarily available to wealthy people. We may not have considered other manifestations of bias. Laura Mandelberg, a Sunrise activist who blogs about fat-visibility and fashion, noted on Twitter that finding sustainably made clothing "is especially tricky for fat people, because most of the 'sustainable,' 'eco-friendly,' etc. brands don't make plus sizes."[8]

Public discourse is littered with spurious comparisons between fat and the climate crisis. No surprise that Bill Maher has called fat-shaming a solution to environmental problems, but even Bill McKibben, from whom many of us learned a systemic, big-levers way to imagine solutions, has used individualist anti-fat rhetoric to illustrate his points. In 2017, he wrote in the *Guardian*: "Maybe it's like all the health warnings that you should eat fewer chips and drink less soda, which, to judge by belt-size, not many of us pay much mind. Until, maybe, you go to the doctor and he says: 'Whoa, you're in trouble.' Not 'keep eating junk and some day you'll be in trouble,' but: 'You're in trouble right now, today. As in, it looks to me like you've already had a small stroke or two.' Hurricanes Harvey and Irma are the equivalent of one of those transient ischaemic attacks—yeah, your face is drooping oddly on the left, but you can continue. Maybe. If you start taking your pills, eating right, exercising, getting your act together."[9]

These are false comparisons. It is true that the Western world has ignored many serious warnings about the climate crisis. And some stroke victims may have ignored warnings from their doctors. But in McKibben's story he's the weirdly glib doctor, while the sick person is both the perpetrator and the victim of this harm, caused by overconsumption. The perpetrators of the climate crisis are CEOs and legislators of various sizes; they are not the primary victims of their own crimes. And some of the most radical climate organizing, particularly on the front lines, is being done by fat activists. Maybe this rhetoric motivates people who are terrified of becoming fat toward climate action, but it does so at the expense of other people's sense of belonging. Additionally, this middle-class conflation of thinness with health distracts from the direct fossil-fueled threats to health—like the ones we discussed in chapter 1. We're proud to be in the movement, but hearing this stuff makes us grit our teeth. In order to connect the dots between the feelings in our bodies—about a dangerous future in a warming world—and the work that needs to be done to avert that future, we must be able to view ourselves, and our bodies, as solutions, not the problem.

When we eat more food than we require to power today's exertions, that surplus of energy is stored as fat in our bodies for future use. It may not always serve us to carry around, but it doesn't harm anyone else if we do. A more apt analogy for body fat might be to sought-after renewable power storage technologies like molten salt or solid-state lithium ion batteries. Besides, every fat-bodied person in the world could undertake a radical weight-loss regime, and the climate crisis would steamroll along unimpeded. And when we understand how greenhouse gas emissions are driven by agricultural subsidies, it's clear that corn, soy, and meat surpluses are not consumer-driven, but industry-profit driven.[10] Blaming fat people for the proliferation of corn-syrup-based products is no different than blaming car commuters for the absence of light rail in their towns. Aside from failing to hold culpable industries accountable, the only thing this accomplishes is disrespecting fat activists and alienating fat people from the climate movement.

We are all made of the struggle to belong where we are, and it's all our responsibility to lessen one another's struggles as we're able. It's not just the loving thing to do, it's how we liberate ourselves. We don't equate health, moral character, or low carbon consumption with thinness. These lies have harmful consequences. Just as there are no "correct" feelings for activism, there are no "correct" bodies. If we succeed or fail, it will be *we* who do so, in these bodies, now.

Practice Being a Body

When we relent in our efforts to constrain, shrink, and control bodies—our own and other people's—then it becomes easier to simply exist in a body. Our bodies are our own spaces, in which we feel our feelings. The most dependable way to do this is through a mind/body practice. This practice can be as pragmatic or as woo-woo as you like. It's anything that brings us into a right-here, right-now, physical experience of ourselves.

Your practice might be anything you do that involves moving and breathing while paying attention. It can be walking, running, singing, playing an instrument, dancing, making love, swimming, raking leaves—as long as you are engaging from your body's perspective. It can be any sport, provided that you listen for your body's communication, rather than forcing it through an ordeal.

Incidentally, most of the activists we spoke with mentioned a regular physical practice as part of their lives. We both depend on body practices as well—Meghan stacks firewood when she's stressed and competes in triathlons, while Josephine practices yoga and takes dance classes.

Why is regularly doing something physical part of a balanced life of climate action?

First of all, because it supports our health. It lets us know how we're feeling, because it causes us to pay attention to our bodies, which is *where our feelings live*. It lets us know how much energy we have—or don't—on a given day, and can help us plan with our own current needs in mind. And while practicing can spend energy, it can also generate energy. It makes us aware of aches and pains and causes us to observe and tend to our injuries. Regular movement helps with insomnia, which can be an unfortunate side effect of climate work. It can help us experience physical joy, which is one of the major reasons that life is worth living.

And it is a cliché to point out that exercise helps relieve stress, but the physiological mechanisms for this are worth understanding. When we experience a stressful event, our body responds to get us through it, with an activation of the sympathetic nervous system, tension in the muscles, adrenaline and cortisol released throughout the body. We then need to move through the experience physically to process or digest it, and return to a baseline.[11] Otherwise, the nerve activation, muscle tightness, and hormones linger, and we keep feeling yesterday's stress, and the day before.

Second, living actively in our bodies helps us develop our emotional intelligence, and shine light into some of the dusty corners where anxiety and avoidance can hide. When strong, negative emotions arise in the body, the head doesn't always want them, but in order to digest them we have to let them reveal themselves and flow. Body awareness makes us less susceptible to the wild mood swings that sustained climate awareness can induce.

To be clear, even when our fear and avoidance is less, we can still be laid low by bad discoveries. Being a body means learning to take the time that we require to digest things that are difficult to endure. We can't serve anyone or anything else if we can't endure the emotional experiences in our own bodies. Through movement and awareness we draw ourselves back down from our minds into our bodies and discover that pleasure and strength live there too, not just fear, fatigue, and shame. By practicing being a body, we can repair a broken link between ourselves and the world around us.

Finally, our body practice keeps us grounded in our sense of human scale. Moving your body around reminds you continually that you have a body—that you *are* a body—which has a size, a location, a life span, needs, and wants. That is, developing your proprioception helps you sustain a sense of your place in the bigger (and smaller) picture. You can depend on this sense of human scale, and of inhabiting your certain place and time, needs and wants, to counteract those spinning-out mind-boggles that overwhelm us when considering the mega crisis. The more we understand ourselves in a physical way, the healthier our sense of limits becomes, of our own selves, of others, and of the world we inhabit.

And related to our physical scale is our sense of temporal scale. Regular practices create ritual landmarks in the day, giving us a somatic sense of time passing, and tools to tell ourselves that work is beginning, or done for the day, even when there's no real end in sight.

A note about media

As we become more aware, we may find that we're more sensitive to the media we consume—especially social media. We use media, we learn from it, we get the word out, or receive word about things that matter to us. But we also feel overwhelmed, outraged, bummed out, sapped of energy, sucked down the infinite scroll. If your feed is mostly swirling ocean-fire vortexes and reports of heat death, you may need to take conscious steps to diversify the accounts you follow and limit your media time. The two of us have always found Twitter to be a particularly volatile place (so much so now that it's unclear if Twitter will still exist by our pubdate). Watching infighting between climate communicators about whether pessimistic or optimistic messaging is accurate or effective can be especially depressing.

A lot of climate messaging is intended to "scare us straight." If this approach doesn't work for you—maybe you're scared already—strategize to avoid accidentally encountering it, whether by having an 8 p.m. cutoff for social media, or by getting climate news from newsletters and digests who speak your language.[12] Or, skip the cheerful "can-do" messaging if that's what gets your hackles up. Before she quit Twitter, writer Kate Schapira observed "No Meta-feelings Mondays" on the platform, in which she joked that she could have feelings, but not feelings about her feelings.

Dr. Sasser came to a similar conclusion when she was interviewing people about their emotional experiences of climate change, which were (unsurprisingly) bad. Her advice was often to "back away from the data, read less, read fewer IPCC reports, read less of the climate science. Climate science is not going to change from day to day. We know what's happening." Instead, she suggests, "focus a bit more on building and strengthening ties within your community, your family. And rediscovering the things that you really love and care about and that matter most to you in this world. And live more closely with those things." Her belief is that for most of us, the earth is one of those things that we love and care about most. Focusing on the things that give our lives meaning helps us stay tethered when big emotions blow through.

Ultimately, once we've found the work we are meant to be doing, and we feel grounded in a human-size perspective, we may find that a little news or info goes a long way. We don't need to be credentialed experts in any subject to be activists. We don't need to be passionate about, or even interested in, the climate crisis to be active against it. As filmmaker and cofounder of Zero Hour Jamie Margolin put it eloquently,

> NO ONE LIKES TALKING ABOUT OR THINKING ABOUT CLIMATE CHANGE. It's like if there was a time bomb about to go off and everyone was ignoring it. And because it endangers everything we hold dear, [some] people are very urgently trying to stop it from going off. And then the people ignoring it say "wow, it's so cool how PASSIONATE and DEDICATED you are to stopping that time bomb. There's this book about it I think you'd like.[13]

FEAR, AND WHAT COMES AFTER FEAR

As we get closer to things that scare us, a helpful reminder is: You will survive looking at this. Fear, dread, apprehension, worry, can all arise in sharp, acute moments and chronic, dull aches. These are normal responses to high levels of uncertainty, and threats to our safety, health, loved ones, imagined futures.

These feelings are present in all of us to some degree. If we don't accept our fear, or learn to make space for it, it will still be present inside our bodies, just repressed and probably leaking out sideways. Acknowledging when we feel fear is necessary to fully live in reality. And naming it can free up a reserve of energy that can be used to scrape oneself off the floor, climb out of the dark, and measure oneself against the task of our lifetime. But if we try to harness and ride that fear toward a goal with no end, we risk fear exhaustion, a possibly permanent philosophical complication from adrenal fatigue.

Over time fear about the future can become like a house-mate. We learn its intimate habits; what time it gets up and goes to sleep, its likes and dislikes. We can grow accustomed to its presence. That original zing is gone. Now when we (oops) go on Twitter before bed, we might read something bad: permafrost thaw, colony collapse, pipeline rupture, or families displaced by wildfire. The news has been sharpened to a tweet and aimed at an imaginary, apathetic reader. We might get anxious, imagining the worst for a restless night. But those spikes of fear always end; we can't sustain them. A political life motivated by fear runs out of fuel. Fear will always be a part of our human experience, and we will live with it, but it isn't the bridge fuel to any habitable future, and luckily we can't feel it forever.

CYNICISM, NIHILISM, AND THE END OF THE WORLD

Fear of change—as a subspecies of fear—deserves its own mention here. Under threat, it's sometimes easier to imagine the world ending cinematically around us than to imagine changing our own lives and the systems that govern them. It's easier to joke about our apocalypse plans than to roll up our sleeves and do the work in front of us here, now, to prevent apocalyptic outcomes. When we reach fear-fatigue but still try to run the same circuitry, misanthropy and cynicism come next. From this vantage point, anything someone else does (from hunching over their phone in the park to eating a plastic-bowl lunch in their car) becomes a perfect example of everything that's fucked up. We'll text "I'm stuck in traffic," not "I *am* traffic." Our empathy suffers, and we may find ourselves fantasizing about leaving society.

But really: why are people's imaginations piqued by surviving the apocalypse and not by the practical steps to avoid it?

Three different threads are tangled here: the worst-case scenario, the apocalyptic fantasy, and a positive vision of the future. Part of the process of finding our core, and returning to it, is familiarizing ourselves with the imaginary futures that live in our minds.

In Lev Kalman and Whitney Horn's 2014 film *L for Leisure*, set during the fictional Long Beach University's 1992–1993 academic year, Professor Sierra Paradise (played by Marianna McLellan) makes a passing remark to a TA for the course Survey of Postapocalyptic Literature: "we underestimate how the postapocalyptic fantasy affects reality, and how its erotic pull might be driving the decisions of our leaders, even unconsciously."

Anthropologist Chris Begley, who teaches survival skills in his free time, told us about some of the apocalyptic fantasies his students arrive with. "There's this idea that you'll go back to a time when, by God, men were appreciated for their skills with an ax and a gun!" he tells us. "You really do get this patriarchal issue. There's also a racial element. You have this idea that 'the country boy will survive!' like the old Hank Williams Jr. song. There's this sense that there's something inferior and unprepared and naïve and childish about city folks. . . . This stuff just comes through when [in] the way people talk about how to defend what you've got against the hordes of unprepared people."

This myth neither begins nor ends with the people in Begley's classes. Bumble, the popular dating app, has a prompt that users can fill out within their profiles: My Zombie Apocalypse Plan. Meghan had a field day with that one, swiping past everyone whose answer prioritized learning to throw axes over making friends. This isn't the Hunger Games. We know the question is supposed to be silly, but the popularity of apocalypse kitsch tattles on the deeply individualistic nature of our fantasies. Ax skills are well and good, but the skill of real value in a crisis is learning how to engage with others, rather than to disengage from (or decapitate) them. In groups, we can weather even massive changes with imagination. Alone, we fear change—and each other.

Whether the threat is zombies, volcanoes, contagion, or Sauron, we've been intuitively trying to prepare for a big challenge by watching and reading. A claim these stories *can* verify is that people understand the value of their lives through their love relationships: however big or technical the threat, the hero's real triumph is reunion and connection with those they love.

For Begley, fantasies of the apocalypse "show us what we want it to be like, and why we don't take action *now*. Because it's going to be this exciting adventure where I get to show off my skills, and

CHAPTER 4

I get to be the hero because I have prepared in these ways, and suddenly I'll be valuable in a way that I wasn't valuable before! So things that you might have been ridiculed for before—you're a prepper—suddenly you're vindicated. Your weirdness is now valuable."

The crucial question this genre doesn't answer for us, however, is: how do we enact our action/adventure fantasies now, with the threat that's *actually* in front of us?

The move here, Begley agrees, is from the individual to the group. Survival is a collaborative act, not a personal one. "Instead of imagining 'what can I do for this particular household? How much canned food can I stockpile?'" he says, "You have to think 'how is this community going to survive? How is it going to work when suddenly there's problems with growing food, or there's problems importing it, or there's an issue with producing the amount of electricity that we do now? How is it that we're going to get through that, and understand that it's not going to be enough to turn your bit of property into a productive sort of mini farm for your family?'"

Zombie Apocalypse Plans can be valid places to work out our existential angst. But it's a sense of not belonging in the present, a rootlessness, *that lets fantasies supplant action in the here and now.* If we're drowning in student debt or barely making it economically, if we are marginalized for any reason, feeling socially disconnected, or we're just watching the mass destruction caused by growth capitalism, the accelerationist fantasy can have some real appeal. If we're not valued in the present, maybe the imagined future will be our day in the sun.

As enticing as the fairy tale is, though, that's not how the story will go: no individual is big enough to hack it on their own. Heading up into the hills to hunt and gather food is not going to be possible for the vast majority of people on the planet. In the world that we *already* inhabit, there are people who are suffering and struggling and not making it—and indeed, there will be more such people in any accelerated version of this crisis. What these fantasies usually exclude is that any existing social problem—hunger, disease, violence, xenophobia—will be worse.

Unless we problem-solve *together*.

GRIEF, RAGE, AND OTHER PEOPLE'S FEELINGS

Climate grief is not one grief. It's many griefs that bump up against each other and compound each other. While we may experience them as one huge painful mess, it can be useful to distinguish among their causes, because the idea that we can grieve for "climate change" is part of the false idea that we need to wear mourning clothes for the rest of our lives. Because, in our lifetimes *the climate will never stop changing*. Grief is a process that has a natural end, and we are able to move through it—however often—in order to live, not just endure, our new reality.

And as ideas like "climate grief" become more common and recognizable (and as they garner more support from institutions like the American Academy of Child and Adolescent Psychiatry), it's important to remember that this particular kind of mental health problem is also a political one, not only an emotional one. Climate grief has an identifiable origin beyond the standard unfurlings of the human mind and heart—it has a series of *political* culprits. The danger in focusing on therapeutic interventions is that powerful institutions will "swap out a political problem for a scientific or technical one" writes Dr. Danielle Carr in the *New York Times*,[14] "it's how, for example . . . climate catastrophe caused by corporate greed becomes a 'heat wave.' For people in power, the reification sleight of hand is very useful because it conveniently abracadabras questions like 'Who caused this thing?' and 'Who benefits?' out of sight. Instead, these symptoms of political struggle and social crisis begin to seem like problems with clear, objective technical solutions." Yes, we need tools for working with and living with climate grief. But it cannot replace political accountability. There is no amount of therapy that can heal the loss of a future.

Similarly, climate rage is not one rage: it reflects a thousand betrayals large and small, a million missed opportunities, countless instances of needless suffering. It reflects inequities and entitlement, the consequences of greed, guile, cowardice, and fear. Here Meghan reflects on the complexities of anger:

> Anger is, unto itself, a reasonable response to things being out of alignment. Anger happens. It's a fair thing to feel, especially when the situation is *unfair*. While I'm not arguing that anger is a cure for everything, I also didn't come here to preach a gospel of "getting over" or "channeling" anger. Anger needs its space sometimes.

> But for women especially, it's always seen as the wrong emotion. Someone is always trying to talk women out of feeling how we feel—and it's so strikingly rare that we're actually asked *why* we feel as we do. Nobody asks! The first response is always to try to police our feelings and make us be quiet, rather than asking us what's wrong.

> That's not okay.

> My own anger is—interestingly—kind of divorced from the actual realities of climate change that I'm apt to experience, myself, as Meghan, as a well-educated white cis woman who now belongs to what's left of the American middle class. I'm not actually afraid of living with much less. I grew up with very little, and I'm a physically tough sonofabitch, and I don't really fear cold or hunger or pain, at least not overmuch. My dad spent the first winter after my parents divorced living in a camper—the kind you prop on the back of a pickup truck, but without the pickup truck beneath it—with hay bales stacked around it for insulation, four of us sleeping abreast on the little foam bed. We spent a subsequent year in an unfinished

basement, which was basically a cement box carved into the ground without the house on top, peeing in a five-gallon bucket and eating soup heated on a camp stove. It was uncomfortable sometimes. But many people live like this all over the world, and it becomes normal enough to some of us.

My anger has very little to do with the discomfort that's likely to be visited upon me. Mostly, it has to do with the fact that these things were preventable, but it felt like nobody cared enough to prevent them.

I suspect I'm not the only one that has trouble disentangling this particular betrayal from others that we've sustained. That is: being worried about a climate apocalypse sits uneasily with being cold at night, especially as a child. The people who are expressions of the problem—in this case, my parents—are not its originators. My parents gave me incredible gifts of connection to the natural world, and to the beauty (and fun!) within it. They taught me to snowshoe and grow carrots and run rivers and hike and care for lilac bushes, and revel in the gentle lap of water on my skin during a dip in the pond. But broadly, adults' behavior around climate at the time I was growing up, and through my young adulthood, was so manifestly tone-deaf! I know that my grown-ups knew that something was wrong, and they felt powerless and it was all they could do to manage the daily reality of keeping their kids fed. I understand that I can't hold them individually responsible for a system's worth of harm into which they were born: not climate change, not poverty, not the late-stage capitalism that produced both ills. Yet somehow it feels like a failure that the generations before didn't fight harder for us–and it feels more that way around climate than around poverty. The irony was, not only were my parents *not* climate deniers, but my mother taught biology and ecology. She actually chose her career in the early 1980s out of a concern for—among other things—climate.

Many years later, when I was no longer poor, when I had been credentialed by some of the most elite institutions in the world, when I was "safe" from a few of the specific scourges of my upbringing, an older scholar I knew hit his moment of existential crisis about climate change. He was a respected social scientist, by no means a climate-denier. Like my mother, he had chosen his field, in part, because he valued ecology and science. He was part of a collective epiphany, among the generation older than us, of people who started to see clearly what the climate crisis means in the lives and hearts of people my age, and who started to take our concerns really seriously. And then they got swept into this sort of grief of their own, which I understand—having been through it myself.

That grief unrolled over him that year in much the same way that it had unrolled over many of his generation around the same time. But then things got weird. There was an odd bearing-down upon me from older people who seemed really committed to me experiencing things *exactly as they were experiencing them*. His grief—not the fact of climate change, but *how he personally felt about it*—became more important than everything else. He decided he would pivot, doing less research and more political organizing. The urgent, panicked need to "do" something seemed to overtake him. So far, so good.

But that rapidly shifted toward some kind of nihilism about young researchers and our careers/lives—just because *he* had finally had his come-to-Jesus moment on climate (conveniently, from within the protected halls of the ivory tower, with tenure and a respected career mostly behind him rather than in front of him), my own right to dream of a career, of research, was suddenly in question. Didn't I understand how urgent it was? Wasn't I concerned for my future on a planet like this?

And here I was, having muddled through these dimensions and feelings and strategies with my peer group, trying to figure out some way to talk about this that brings us some clarity and some peace, and with them I had found the ability to get up every day and be generative and creative. I work really hard; I'm a faculty member and I represent my community in the Rhode Island State Senate, and it's a second full-time job that's only minimally compensated. It should be a full-time gig, but I need to earn a living and so I, like many legislators, do double duty. And then along comes this fellow wringing his hands: "Oh, it's so terrible! How can you even think about writing academic papers anymore? When the world's on fire?" Says the senior academic! Who already had his chance! Of course I am tired. But I'm also tired of consistently being told that whatever I am feeling should *instead be what somebody else is feeling*—in this case, what *he* was feeling. The climate crisis has been with me for literally my entire life! I promise I haven't somehow missed out on the fear and discomfort.

And also: don't mansplain grief to me, bruh.

These kinds of misunderstandings—born of radically different experiences of the same phenomenon—happen a lot around tender subjects. Dr. Matthew Goldberg, an associate research scientist at the Yale Program on Climate Change Communication, reflects on how other people's feelings often show up irresponsibly: "I think that the knee-jerk stuff comes from grief. I think sometimes it'll come from shame. But frequently, it comes from grief. Sometimes it comes from surprise . . . so again, I think the impulse is often to minimize or push aside." People also frequently use ridicule as a first line of defense against others' grief, or shame, especially when they're older and/or in a position of greater power. This makes dismissing young women's feelings particularly easy.

Grief and rage happen—and other feelings happen. They are natural responses, and they arise for good reason. The point is this: we can't slough our own pain off on others, especially on those younger than us. That's entitlement. In the next chapter we talk more about generational rifts and how we can heal them, but for now our own responsibility is to live honorably from our own core, and deal with our own feelings so others don't have to deal overmuch with ours.

A FIELD GUIDE TO DENIALS
Denial with a capital D

The villain in climate crisis mythology is the "climate change denier." They are also known as the climate skeptic, contrarian, or obstructionist. This could be a loudmouth in your family, class, or congregation. They make up a disproportionate percentage of legislators. We may feel powerful—or anyway our outrage grows strong—when we attack this person's specious arguments. Climate activists have used public debate with these liars as a way to expose more people to facts about climate change, and as a way to ventilate some of the aggression that builds up doing this painful, difficult work.

But the debate has been over. Scientists debunked the last scraps of denialist "evidence" thoroughly years ago. Denial among people in power is only a commodity that fossil-fuel companies are trading to prolong their profits. It is no more useful to argue with an oil-backed legislator than it is to pump quarters into a pinball game. The house gets the money no matter how well you play. It's important to remember that if this crew quit and went home, while there's money to be made doing so others will assume those roles. It matters less who they are as individuals than who we are.

Big-D climate deniers are a small group, and we can outorganize them. When we know that an adversary argues in bad faith, the only course of action is deplatforming— preventing the spread of false information by blocking Deniers' access to communication platforms—not debate. And as for arguing with a bellicose relative, only you can judge the merits of such an undertaking.

Our own real work is on the subtler forms of denial.

Small "d" denial

There are two other kinds of denial that don't get much discussed, but are critical forces in shaping what's possible. The first is a progressive soft denial; we say climate change exists, or we "believe" in it, "science is real," and then we change the subject. We establish our membership on the good team, and then push it from our minds.

Climate change is a distinctive issue because it's this existential thing. Everyone is hardwired to want to deny it. Everyone has emotional defenses against it. We do, too. It's not a partisan issue to want to protect yourself from thinking about really depressing shit all the time.

This is pervasive among Democratic and Republican politicians, progressive activists, and well-intentioned citizens playing along at home. We are struck daily by how many thoughtful people still have a huge mental block around the climate crisis. It can take the form of an emotional performance of how bad you know it is, a cynical joke about how "we'll all be underwater soon anyhow," or silence when you might instead have mentioned the issue. But whatever its verbal expression, this form of denial is always accompanied by inaction. We've talked about letting the mind recognize what the body is experiencing. Dr. Cornel West describes an inverse state of affairs that he calls *moral constipation*. This is when the mind knows the difference between right and wrong, but won't let that information flow into feeling—into the body—because it might mean we'd have to change.

Small "d" denial is understandable: climate change is terrifying. The human psyche doesn't do well with ever-present existential threats. Self-protection is instinctive, and if you acknowledge the full depth of the crisis, inevitably your life must change. So as would-be allies practice soft denial, the people doing the work shoulder a much heavier burden. They carry the full weight of

CHAPTER 4

the job itself and the growing weight of bad odds, made worse by lack of support. When the reality we observe is deemed extreme and unrealistic, and the necessary solutions are ignored or called politically impossible, we struggle to keep moving forward. If you find reassurance in the idea that other people have climate activism under control so you don't have to think about it, today is a great day to make a date to talk with one of those people and ask how they're doing.

Healing denial

But here's the thing: the atmosphere doesn't care what we're thinking about, or feeling. We are allowed to spend many hours every day *not* thinking about ecological catastrophe. What matters to solve the problem is *what we do*, not how we feel. This is where we encounter the other form of denial, one that we've found is essential to our well-being. It acts as a reprieve, a healing rest: a periodic window of time arranged by our psyches to lay down the burden of knowledge as we get used to the weight of it. We don't really forget what we know, but we can exist for a time as if it weren't so.

We enjoy an unseasonably warm day in January. We fantasize about what we'll wear to our someday-grandkid's wedding (or whatever works for you). We watch *Star Trek* and embrace its simple humanistic optimism. This denial doesn't get in the way of doing the work; it's a natural and necessary coping strategy that restores some of our capacity to handle the work.

The downside to this strategy is the inevitable remembering. And really, the mechanism evolved to heal singular, temporally specific traumas and losses; a loved one dies, a home burns down, a lover leaves, you are robbed, or assaulted. Time passes, and the source of the pain moves deeper into the past. Your physiology does what it can to digest the pain, and your mind makes use of the perspective time offers. It's never the same, but completeness, doneness, helps a lot in healing. Even in a war or a disaster, the fall of a nation or life-support system, there might be some peace to be had in the knowledge that it's happened before to others, that it is certain to end, that humans have endured it before, and you can read the stories they lived to tell. We don't have that assurance now, of a cycle that repeats; we are watching a cyclical system unravel. But we do have the tools, skills, wisdom of our ancestors, as well as each other, to depend on when the going gets unbearably tough.

DESPAIR, DEPRESSION, AND RETURNING TO THE CORE

For all of us there are times when the work becomes impossible. When we've spent our inner resources, and fear tips over into despair, or grief gets locked up in depression. Sometimes depression can feel more like boredom than despair. Boredom and frustration—with ourselves or other people—can overwhelm us.

We can call this burnout, which is absolutely assured when you are tasked with caring and investing indefinitely, with unceasing demands and unmeetable expectations. But in the context of climate work we find it helpful to think of it as the downswing, because it is a reliable, predictable part of engagement. When we experience the inevitable downswing, it's time to return to the core. What we're cultivating at our core is a personal relationship with what makes life worth living.

This means we savor joy where we can, and pleasure where we can. This means we seek catharsis, not because it enables us to be better activists, but because joy and pleasure are core parts of being human, our human right. This means we play for sheer love of it. This means we rest. We believe that if a climate activist has to quit their campaign for a month or a year to remember why it's amazing to be alive, that is the next right step. If they find they need to change their kind of engagement completely in order to stay present, that's the necessary work.

We're not "preventing burnout" in order to instrumentalize ourselves as workers, or send ourselves back into the trenches as cannon fodder, but because the hope and energy in each of us is a precious resource that animates all parts of our life. It can flow in nourishing abundance and it can be depleted, damaged, snuffed out permanently. This resource is in our relationship to ourselves. It is our starting point, ending point, and the basis for everything in between.

We've inflicted a long chapter on you, and we're about done. For now, though, we want to emphasize that we all have all these big feelings—it's how we live with them that's important. Even in the midst of a total chemical makeover of the world, and a liquidation of species from our shy forest cousins to the unnamed movers and shakers in the soil, it can only make sense to understand our lives in the terms of our lives: who do we want to be, together? What would make us glad to live, instead of simply afraid to die? And if we find answers to these questions, can we not instrumentalize them immediately, but simply feel them first?

After years of emotional climate work, our belief is that people don't want to hear about the harm we're implicated in because they fear that then they'll have to give up everything and feel horrible forever. A call to "climate action" alone can't speak to this fear. But we know that after we grieve there is a shift, and it comes like the peal of a bell, the dropping of a burden. We begin again as we mean to go on, and as we're able. We'll continue to plant together, whatever does or doesn't bloom. We can live in the world we want to create. We can embody the best of what we're fighting to protect. But in our own hearts all we can do, all we need to try, is to stay in love with this world as it dies, and fall in love with the next one as we bring it into being.

CHAPTER 5

Family Planning

The Labor of Love

The humanist part of me thinks of the word "karass" when I think of family. Coined by Kurt Vonnegut in Cat's Cradle, *a "karass" is a Bokononist (a fictional religion in the book) term for a "group of people brought together to do God's work—though the purpose of that work is not something they can ever be fully aware of." I don't think any one of us have any idea what our purpose on this planet is, except probably to fart around, but I do know that we've been brought together under peculiar circumstances.*

—Ben Koch

FAMILIES ARE MADE OF LOVE, NOT BLOOD

The phrase "family planning" summons contraception to mind for many. We want to propose here that family planning can mean a lot more, too: planning *to have a family of any description*, planning on behalf of your family, planning as a part of a family. The climate emergency has become a factor in all of these plans.

While we are trying to stop the heating of the planet, we also recognize that there is no lever big enough to—geophysically—revert our climate to the mid-nineteenth century. Climate disruption is something we will always live with; unlike a war, climate change will not end. In difficult times, and for the long haul, our families can be great sources of fortitude and joy.

For those of us who experience family as a place of conflict, anxiety, or estrangement, this may sound irrelevant or nightmarish, but here we want to be clear: families are made of love, not necessarily blood. They're defined by long-term commitment, and often by intergenerational commitment; by a sense of belonging and being in it together. One of our respondents told us that to her, family was "people who I want to keep seeing into old age." We like the following clunky but comprehensive definition: "a family consists of any combination of two or more people, bound together over time, by ties of mutual consent and/or birth, adoption or placement, and who take responsibility for various activities of daily living, including love."[1] Families are people who care for each other's lives.

And there are some things that family does best. We can get science and policy education elsewhere, but our parents and elders are who we turn to with moral questions about big issues, and the younger generation is often where we direct our most precious gifts. Through family we get to be essential links between the past and future, and essential bonds within our communities.

A lot of the struggles we may have with the idea of family actually arise from its limited definition as *nuclear* family. This is particularly true because of how the nuclear structure—the idea that two parents and their children make a family—is pushed as a "solution" to a made-up problem, rather than understood as a brief departure from a global, historical, human understanding of kinship. And this departure has caused as many problems as it has solved. Over the past thirty years we've all witnessed hand-wringing about the "breakdown of the nuclear family"; US policy nudges people toward marriage with the promise of lower tax rates, favorable insurance premiums, and legal recognition. The nuclear family-or-bust agenda fails us by valuing that small unit to the exclusion of all other structures. And it remains the political ideal, despite the fact that less than a third of Americans, and less than half of US kids, currently live in that arrangement.[2]

We reject that limiting definition of nuclear-family-as-the-only-family, but it's clear that, regardless of the legal or social definition of these relationships, everyone needs family. People only thrive in love—love for and from others. And families make us stronger in measurable ways.

Let's start with marriage, which is, medically speaking, a "privileged health status." Data shows that married adults have better mental, emotional, physical, and cognitive health, and typically live longer than unmarried ones.[3] Furthermore, the overall health of never-married adults has worsened over time.[4] The COVID-19 pandemic revealed this clearly: the 2020 lockdown took an extra-heavy toll on single adults, who were coping with illness and uncertainty alone.[5] Happy marriages confer social and health advantages.[6] But marriage—as a legal status—bestows its benefits unevenly: married people are more likely to be white and wealthy, especially in the United States. Marriage also carries plenty of gendered baggage that many are eager to discard. And it is limited by law to two people.

Marriage is a privileged status for other, non-health reasons as well: married adults can visit each other in hospitals, inherit each other's estates, share legal guardianship of children. How we define "family" shapes social policy like parental leave and childcare tax credits. Caregiver benefits, family medical leave policies, and the like are also tied to legal family status. All of those privileges make life easier. And while "marriage" falls short of defining the many ways people couple up and/or form families, researchers tend to use marriage as an easily studiable shorthand for shared domesticity. And the research picture, incomplete though it is, underscores that going it alone is a struggle, especially when the going gets tough.

Data also shows that crises (including, for instance, pandemics) have the potential to *increase* the social importance of families for the same reasons, *as well as* widening existing disparities between familied and un-familied people.[7] In other words: family, for all its foibles, operates as a buffer against many of the things that go wrong in the world.

So the questions we aim to answer in this chapter are: how can we grow family resilience? How can we talk about the climate emergency in our families so that it's a point of connection, not conflict and anxiety? And how can we cultivate kinship in all our relationships? In this crisis, all forms of family planning become urgent. For us, that urgency means our only priority is love, in all its forms. This is an invitation for each of us to get comfortable with our many relations, and to think creatively about how we love them.

PLANNING FAMILIES

The word *khaandaan* in Urdu means "extended family"; Urdu also has a word that refers to a close group of family and neighbors— people who share a fate in an immediate geographical sense (the word is *kunba*). In Wolof, as in Urdu, there is no word for "nuclear family"—the word often used means "relative": *mbokk*. The root of *mbokk* is the verb *bokk*, which means "to share" (e.g., Josephine and Meghan can *bokk* a bowl of rice), as well as "to belong."

There are as many ways of describing family as there are cultures and communities, and many of them focus on a shared sense of belonging. "For the Ilongot people of the Philippines, people who migrated somewhere together are kin," David Brooks writes in an essay called "The Nuclear Family was a Mistake." "For the New Guineans of the Nebilyer Valley, kinship is created by sharing grease—the life force found in mother's milk or sweet potatoes. The Chuukese people in Micronesia have a saying: 'My sibling from the same canoe'; if two people survive a dangerous trial at sea, then they become kin. On the Alaskan North Slope, the Inupiat name their children after dead people, and those children are considered members of their namesake's family."[8]

The point is that humans have a global history of inclusive, loving relationships with each other: the political arrangements of the nuclear family are recent, and in many ways limiting, phenomena.

The structure of how people in the United States live together in family relationships is changing. In 2021, a full quarter of US adults ages 25 to 34 resided in a multigenerational family household, up from 9 percent in 1971 and—as of 2023—the number was still rising.[9] In other words, despite the "2.5-children-and-the-white-picket-fence" trope, the reality for most of us is different. This shift is reflected in the reemergence of phenomena like the "in-law apartment" and the "granny flat" or the real estate moniker of "two homes under one roof." These are architectural representations of changing family shapes.

Especially as Americans cope with rising student debt and housing costs, multigenerational living can mean lower expenses and increased savings. Data shows that people are living multi-generationally for financial reasons, especially if they're relatively recent immigrants, but that there are emotional benefits, too.[10]

Multigenerational (or otherwise extended) families confer many blessings, but two in particular stand out to us. The first is resilience: there are more people around to lend a hand, pick up the slack, or to share care and domestic work. People living together inter-generationally have lower rates of depression than their solo neighbors, and children get cognitive and developmental benefits from living in multigenerational homes.[11] Extended families have more people to shoulder unexpected burdens—when a kid gets sick during the school day or when an adult unexpectedly loses a job or when a teenage child is feuding with a parent and needs a different adult hand at the shore. A nuclear family, by contrast, puts an awful lot of pressure on two adults. There is no slack. There is no shock absorption. Rev. Mariama White-Hammond, whom we met in Part I and who does not have children, often cares for nieces, nephews, and other children in her family. She says,

I do think *not* having children in your life is a loss. So many children that could use extra support, so many families that need more help. The way people are trying to do this . . . how many people do I know who are single parents? How many parents do I know that both of them are working so many hours they're struggling? For me [not being a parent] is not a loss—it's an opportunity! It's an opportunity for us to care for the kids that are here. [To] give them everything that they deserve, and to have adults who are actually able to double down and provide extra support.

Greg Marshall, a climate activist and member of a three-person polyamorous partnership, concurs: "Being in a little 3-person V," he says, "feels much more dynamic than the monogamous relationships I've been in. When one of us is sick or exhausted or out of town or unavailable for whatever reason, there's still two people who can walk the dog, do the dishes, and keep each other company." He smiles. "Having all these perspectives around me keeps me humble about what I know and what I don't, and keeps me in a state of mind where I want to learn. Having so many caring relationships around me makes me feel whole, which makes me better as I apply myself to things out in the world."

Many hands, in other words, make light work.

The second great strength of big families is that they are a moral unit, stronger for their size. Multiple adults teach children right from wrong, how we behave toward others, and how to be kind from different perspectives. Kids assimilate more viewpoints, becoming the interpersonal equivalent of multilingual. There are more love relationships for both adults and children, and in large or extended families, roles can be less restrictive for everyone. Multigenerational families can draw upon the wisdom of people at many points in the life cycle: very old people, very young ones, and everyone in between. There are many people to share your gifts with, to learn from, and to teach.

The thing is, neither of these blessings is inherently conferred by genetics.

"At the end of the day, I feel like legacy and family is not restricted to blood," says Rev. White-Hammond. "I say that because I believe that a lot of the challenges that we face in climate change are also connected to an individualistic, small-thinking way of living that says 'I'm responsible for myself and this particular family unit!'" This, for her, has big political consequences: "our ability to limit who we are directly responsible for and who is in our world actually *prevents* us from having the kind of empathy that we need on a much broader scale."

Additionally, the people who can make the nuclear family work frequently rely on wealth to pull it off. Conservative commentator David Brooks described this dynamic with surprising clarity,

> Think of all the child-rearing labor affluent parents now buy that used to be done by extended kin: babysitting, professional child care, tutoring, coaching, therapy, expensive after-school programs. . . . These expensive tools and services not only support children's development and help prepare them to compete in the meritocracy; by reducing stress and time commitments for parents, they preserve the amity of marriage. Affluent conservatives often pat themselves on the back for having stable nuclear families. They preach that everybody else should build stable families too. But then they ignore one of the main reasons their own families are stable: They can afford to purchase the support that extended family used to provide.[12]

There is a class-based possessiveness to privileging the nuclear family, or the blood family. In other words: the bigger the estate, the less generous people are with family titles. Amassed wealth can make the mutual obligation of social networks irrelevant, as the focus shifts to protecting your goodies for your heirs. The political phenomenon of the nuclear family has become a very effective excuse for *not* being responsible for anyone beyond "our own." That's a political outcome. By expanding how we think about family, we also expand who we are responsible for, and accountable to.

In the 1950s and 1960s the nuclear family became the cultural ideal, supported by policy. There were a few things that made that possible at the time, including: earnings (one person working outside the home could usually make enough to support a family), as well as economic opportunity. Women were also legally denied entry to the workforce in all but very specific circumstances, which protected both men's social status and their jobs. Beginning in the 1970s, the economics of family life changed radically—real wages decreased, women and people of color fought for and won some significant legal protections, and neoliberal policies, including privatization and decreased social protections, began to take hold.[13] The point here is that the *Leave It to Beaver* version of "family" was a blip on the screen of history—restricted to certain races and classes of people—rather than an enduring feature of human relationships. It had very specific political and economic enablers that no longer exist.

Rev. White-Hammond told us that an expansive definition of family is part of her cultural identity, that has a root in her ancestors' experiences of slavery:

> It is particularly part of the Black faith tradition, and I think some piece of this is because Black families have gotten disrupted since slavery. There's been all sorts of reasons, like over-incarceration. All sorts of reasons that Black families have been disrupted. And so maybe that is part of why, from the time that people landed on these shores, we're like "we're not even from the same tribe? Whatever, we gotta figure this out!"

We in the United States are clearly at a moment when the meaning of family is evolving, by necessity as well as by choice. As the climate emergency alters our plans and demands more of us all, we can learn from each others' experiences. We can all imagine bigger, now, and plan for more mutuality, more resilience, more loving family structures. So we can look to people who are living, and have lived, outside the nuclear convention for more expansive visions of family.

Queer families

It's a common account within LGBTQ+ communities: since many queer people are rejected (or simply tolerated) by their families of origin, queer communities have often led the reimagining and redefinition of contemporary families.

Eli Pushkarewicz is a white trans/nonbinary person who works in disaster relief, and who we encountered briefly at the beginning of the book. Their snapping black eyes set off their mischievous grin, but despite their exuberant good humor they can turn instantly serious. They tell us that family is very important to their primary partner Annie, so although they decided not to have biological children, they have chosen to foster—and they foster primarily older children whose needs are different from those of infants or toddlers. "A significant number of our chosen family love children but do not want them themselves," Eli tells us. "They have been very excited to help us nurture the kids that come into our care, since we don't know what the needs are/types of youth until we get a short-notice call that they are about to be placed."

As part of the process to be approved as foster care providers, caseworkers wanted Eli and Annie to be intentional about how they introduce kids to the "family." The couple chose to put up a wall of pictures with chosen family from around the globe, to prompt conversation about who will be present in children's lives while they are staying with the couple. But in order to foster at all, Eli and Annie still needed to conform to some of the mandates of the institutional mainstream. Eli continues,

In order to foster, my partner and I had to get legally married, which is a requirement for any couples wishing to foster or adopt in Maryland. We had a commitment ceremony years prior with our chosen family, but marriage was not then legal in our state. In the time post-Supreme-Court-ruling-making-it-legal, we had not filed the official paperwork, as I strongly believe legal marriage is an antiquated construct meant to prioritize heteronormative families and restrict resources and legal protections from alternative families.

However, if they wanted to foster, they had to comply. This made them consider how to engage with their chosen family intentionally:

For those who have biological children, [raising a family] starts with a baby shower. And there are lots of societally-recognized ways family, close friends, and coworkers get notified that family dynamics are about to change, and how the changing family may need their support.

How would Eli and Annie signal their impending parenthood? They had no baby shower, no collective ritual designed for a situation like theirs. As they prepared for fostering and considered what it would mean to have a succession of children coming through their home temporarily, and at short notice, they decided they needed to explicitly call in the reinforcements:

When we were hitting our last phase of the foster screening process, we decided to have a party. It was a barbeque, with the purpose of introducing the concept to our chosen family, getting them excited, and also empowering them to self-identify and commit to how they wanted to support us. Also since they came from different areas of our lives, it introduced the different groups to each other. One of our Buddhist friends performed a ritual where our chosen family talked about their commitment to helping us build our family, their wishes for us.

Eli and Annie created an event that both declared and helped form a bigger, more interconnected family. It combined familiar elements (barbeque, celebration), with new ideas and introductions, as well as a specific spiritual aspect. They took traditions that were already meaningful—and fun—to them and their community, recombining them to make room for new relationships. This is family planning in action.

Poly families

Poly communities have also had to be at the forefront of imagining and redefining families, because the institutions of lives—ranging from marriages to mortgages—are designed for two partners, rather than multiple adults.

Polyamorous families are families in which adults have more than one romantic partner, with the knowledge and consent of all involved. We spoke with a "polycule" of three in Rhode Island: Courtney Klein, Ben Koch, and Greg Marshall (these folks have asked that their real names not be used, so we've given them pseudonyms). Courtney and Ben had been romantic partners for about six years before Greg joined their family.

"Greg has been in our life for the last three years," Ben tells us, "but I have a hard time remembering life without him." Ben and Greg are both partnered with Courtney, but not with each other, forming what they call a "V." "It's important to note," Ben continues, "that shortly after Courtney and I opened up our relationship, I went through a very serious medical crisis and needed all the support I could get. But I ultimately could not give much, if any, support back. The timing of Greg's arrival into our lives was all too perfect, as he could help support Courtney who, at the time, was busy keeping me alive. As my health has recovered, our 'V' has flourished."

Greg tells us of his journey into polyamory thoughtfully. "I think templates are really helpful for lots of people. For some people well-established ways of being help keep them stable.

That's good and healthy—I'm all for couples who find their true home in cis-hetero-monogamous relationships." He pauses for a minute, considering. Greg is a white man in his late thirties, with an overgrown haircut and black skinny jeans. He continues,

> The problem is when social templates become one-size-fits-all social expectations. I grew up in a neighborhood surrounded by a lot of "normal" monogamous middle-class heterosexual married couples. Everyone had tied their knot and all the knots seemed to be holding just fine. I figured I'd certainly tie mine at some point. I was uncomfortable with it but didn't know how to voice that discomfort without coming across as greedy or cavalier. I felt like I had to turn a part of myself off.

For Greg, polyamory offered a way to define romantic relationships *positively* rather than *negatively*. The measure of whether he was good to a partner wasn't defined in terms of rejecting all others. "I found," he says, "a different kind of fidelity that was defined more by how I showed up for partners: what energy and ideas I brought to our interactions, the gifts I gave, the care I offered." Greg and his partners still have boundaries, but getting out of the monogamous template has forced them to actively define those boundaries, to "know each other rather than know a mold. It's made me a better partner and makes me feel more whole."

Courtney, Greg, and Ben are a family. But as with Eli and Annie, some institutions recognize their family, and others do not. Ben says, "I don't believe there is enough recognition for our family. A partnership between Courtney and myself is very much recognized, but as soon as you include a third into the equation, all the wheels come loose." He takes a breath. "It'll be very interesting to see what happens when we decide to purchase a house, apply for a loan, get married, etc. Will one of us just stand aside for legal purposes?"

Greg is thinking about these things too. The three of them cannot share a healthcare plan, which is both expensive and frustrating. If two of their polycule enter into a legal relationship and the third ends up incapacitated, will the two be able to make decisions on the third's behalf? But, he says, the question of parenting as a group remains the biggest challenge: "My biggest concern is child custody," he tells us. "If we decide to have a child and raise them in a three-parent household, will we all be recognized as parents by schools, doctors, or the state? It's hard to imagine all the potential hurdles. What about the nightmare scenario where the two legal parents die in a car crash?"

In 2020, the city of Somerville, Massachusetts, broadened the definition of "domestic partnership" to include relationships between three or more adults, expanding access to health care. The law only applies to city employees and it remains to be seen whether private employees will follow suit.[14] Regardless, Somerville is the first and only municipality in the country with an ordinance like this.

Even as white professionals, with the racial and occupational privilege that implies, Ben, Greg, and Courtney, are cautious about talking openly about their relationships, although they are not closeted. Courtney tells us,

> I'm extremely grateful for (and intentionally chose to live in) a city with a thriving queer and trans community who have paved the way for a culture of broad tolerance and acceptance when it comes to love. I sometimes feel like we—as a white, cis, straight-ish polycule—are riding on the coattails of their hard work and hardship.

Aunt(ie)s

Meghan and Josephine are both aunts to biological, adopted, and legally recognized relations, as well as to other niblings whose kinships are not acknowledged by the legal system. We use our own example to open a conversation about the big, mutable role, often called Aunt or Auntie, that can reflect all kinds of core familial relationships based on love and responsibility and accountability. Meghan explains:

> A few of my niblings call me *Tia*—those for whom Spanish is the first language. "Tia" means "aunt" but it sits, linguistically, somewhere between "aunt" and "auntie"; it's applied to blood and beyond blood. The niblings for whom English is a first language call me by my first name: Meghan, or some variation of that (often Meggie). I love Meggie for the same reason, I think, that I love Tia: they reflect something that feels special.

Rev. White-Hammond describes a related experience:

> I remember one person, a friend of mine who was white was like "how many family members do you really have?" and I was like "oh, well you know probably 50 percent of the people you hear me refer to as Aunt and Uncle are not biologically related to me and that's because I grew up in a church community where you didn't call any adult by their first name, you called them Aunt and Uncle or Brother and Sister."

In India, the words "Auntie" and "Uncle" are traditional terms of respect for older people who are not strictly blood relatives; by some accounts, the use of "Auntie" and "Uncle" was introduced in the Caribbean by indentured Indian laborers who brought the tradition across the ocean with them. In 2017, after US Congresswoman Maxine Waters dealt a series of verbal takedowns to Donald Trump, she was monikered "Auntie Maxine" by young people on social media.

While Waters embraced the nickname, other prominent Black women in the United States did not (including Oprah Winfrey, Mary J. Blige, and Nikole Hannah-Jones), because of the word's association with slavery and Jim Crow. "I am interested in what it means to name a social role for an older woman who is seen as liberated and not defined by caretaking," writes Imani Perry of the Maxine Waters incident. She continues,

> Reels on IG and posts on TikTok celebrate aunties. They are the women who come into family gatherings unfettered and free to express themselves fully. They may or may not be in the kitchen. They claim their own identities. They do not require a husband or children to be valued, though they may have both. They are shown dancing, playing cards, and telling stories. The depiction, for me, feels freeing precisely because according to cultural norms, Black women of a certain age are not expected to prioritize their own interests, desires, and well-being. "Aunties" don't play that. Aunties are fun. Aunties are playful. Aunties give advice with candor, something mothers often hesitate to do, because they are preserving the illusion of parental respectability.[15]

What stands out to us here is the exploration of the idea of "Aunt" or "Auntie," as a name for a woman who lives on her own terms, and who still has responsibilities and love-ties to family, but who is less bound by the narratives of self-sacrifice. What is equally interesting is the adjacent absence—the word "Uncle" is comparatively cut and dried. We don't have an affectionate "Unclie." There is less generous chosen-family use of the term in this country, and perhaps fewer preconceptions about the affective relationship a child might have with their Uncle.

Stepfamilies and blended families
When people ask how many siblings Meghan has, she often says "it depends on how you count." She elaborates,

> It's an intentionally equivocal response, but I'm trying to make the point that my family is big and expansive and some of those relations are recognized by law and some are not. And they are all my family. As a child I had a stepmother, a stepfather, each

of my parents had a handful of live-in partners, as well as two stepsiblings. As an adult I had another stepmother, three more stepsiblings later on—you get the picture. My niblings—all eleven of them—joined the family in different ways: through adoption, birth, marriage, love. They're all family. The words themselves that I'm asked to use are cumbersome and too fastidious. They do not reflect my experience living in my family.

The US Census Bureau estimates that 1,300 new stepfamilies are forming every day. Stepfamilies or blended families, like all families, take many forms. Children may reside with one parent and stepparent and visit another; they may move back and forth between parents' houses; they may live full-time with a parent and stepparent, as well as stepsiblings. And this is a reality for half of the young children in the United States: 50 percent of the sixty million children under the age of thirteen are currently living with one biological parent and that parent's current partner.[16]

For us the point is this: the sheer number of stepfamilies and blended families stands in stark contrast with the idealization of enduring heterosexual marriage and biological children. Chance and circumstance define all our lives as powerfully as values, geography, and laws. When we have people to love and relationships that sustain us, those are families that deserve respect and recognition, whatever their configuration.

BEFORE, DURING, AND AFTER: GENERATIONAL EXPERIENCES OF CLIMATE CHANGE

Climate change serves us with at least two big crises: the geophysical crisis (how the planet is heating, and what the consequences are), and a societal crisis (how we experience and respond to this chaotic transformation). All these things play out in the context of our families, and while families can be the first source of information, strength, and comfort for many of us, they can equally be the source of conflict. Fears and resentments are playing out intergenerationally in families and beyond.

As the temperature rises, so do tempers. We see an increase in finger-pointing, guilt, and blame, including across generational lines. But the story is more complex than "the greedy boomers" or "the lazy über entitled millennials" or "those Gen Z snowflakes." Juan-Pablo Velez, an organizer we'll talk more with later in the book, had some choice words to illustrate the conundrum:

> In 2020 of all years, I finally decided I wanted to have children. My immediate next thought was: what am I going to say when it's 2035, and my bratty kid looks around and sees everything on fire and then naturally blames me? As someone who's been so angry at the adults in my life for not taking action, I see it so fucking clearly that Generation Alpha and beyond are gonna blame *us*, because we're going to be the old people. And it just seems intolerably unjust because, if anyone, *we* were the ones who had no power! This is really going to open up in the future: a lot of the obstructionists now won't be here. And we'll be left holding the bag with a bunch of new people who we chose to bring into the world, holding us accountable.

This rift needs healing in all directions.

The thing is, any generational failures that adults don't address explicitly *now* will keep on rolling downhill. It's our ethical responsibility to handle the emotional dimension of the climate crisis, at the very least; at the very least, we can try to interrupt intergenerational trauma along these lines. Because unless we develop *some* tools to communicate between generations about climate change and what it means to each of us, we hand off an emotional legacy of silence and blame (in addition to an ecological legacy of climate harms) to our kids. That's hardly the gift that anyone wants to be remembered for. And as we described in chapter 4, the emotional burden also makes organizing hurdles that much higher: action is almost impossible if you're too jammed up with your own pain.

Learning to talk about the climate crisis, we've found, is a lot like learning to talk about sex: uncomfortable and vulnerable at first, easier with more skills and more practice, and crucial for any fulfilled and healthy relationship. Ideally, conversations about sex are about how consenting parties come together to do something intimate: about how to bring pleasure and to avoid harm, how to explore what feels good; how to balance the needs and wants of the people involved; how to honor boundaries; about creating some shared understanding of what you're up to and why. There's some technical mastery involved too, like figuring out contraceptives or PrEP, and there's a question of how the technical tools like contraceptives or PrEP interact with your body, or your partners.' Climate, we think, is a lot like this: a climate-changed world is a new one that we all need to learn to navigate on shared terms. We can learn, in this context, how to bring pleasure and avoid harm and explore what feels good; we can learn to balance the needs and wants of those involved, define boundaries, create a shared understanding. We can confront our own insecurities—both individual and collective—without expecting that someone else will guess how we're feeling. We can be responsible for ourselves, and responsive to others.

Getting used to talking about it helps, and helps unlock our imaginations. And we really, really need our imaginations now, because there are no prefab options for the work that we're confronting. In this sense—and extending our sex metaphor—the communities who often have the best ideas are the communities who aren't mainstream, for whom the prefab options have *never* fit, and who have had to flex their imaginations much more to build resilience and reduce harm. Queer communities, poly communities, kink communities: these are all groups that have had to develop skills for talking about sex much more explicitly than in the mainstream. If you were a gay man in the 1980s, *your literal life* (and those of your partners) was on the line when you communicated about HIV. And so too for people of all ages in a climate-changed world: our very survival depends on our ability to talk and listen to each other. The layers of guilt, shame, and silence are compounded, and that won't get better unless we discuss it.

The good news is that communication is a skill that can be developed.

Although the global temperature began rising generations ago, and the climate will be changing for generations to come, there is a specific window of time that divides our experiences in a meaningful way. It begins with the consciousness that climate change threatens our lives: assume a variable start date for different people, but generally people born in 1980 and after came of age to this awareness. We can define the closing of this window politically as the deadlines to meet 1.5- or 2-degree targets, and personally as the end of our reproductive years.

When we're talking about intergenerational love—or conflict—in a changing climate, we have found that people's experiences fall into three rough groups relative to this window:

- people who made reproductive decisions—and more broadly, major life decisions—*before* this window: whether we are related by love, blood, or simply shared humanity: these are our Elders

- people who are making reproductive decisions and major life decisions *during* this window: this is us

- people who will make these decisions *after* this window has closed. No matter whether or not we become parents, these are all our kids.

Us

There is a particular agony for those of us who are planning families now: if we could just *see through that window* . . . but our generation is defining itself through impossible decisions made without a clear vision of the future. Sometimes this unique hardship can make us furious. Some of us were children when James Hansen testified about global warming to the joint houses of Congress in 1988. Some of us were not even alive then. In school we celebrated Earth Day, picked up litter and learned about endangered animals, while emissions climbed and rainforests burned. But if we are angry at our elders for not preventing this harm, we can imagine how our children will feel with whatever we leave to them. If we have lost faith in our parents' generation, how do we keep faith with that of our kids? Even if we decide not to parent, are we not still morally accountable to those that come after us?

The biggest generational gap is visible in the knowledge that global warming will pose a serious threat in one's own lifetime.[17] This doesn't mean that no older people have felt climate change's effects; it just means that they're not going to have the same experience of it as a child born in 2013. Dr. Matthew Goldberg, who we met in the last chapter, notes: "Younger folks are more engaged, more worried about climate change." He pauses. "But that's almost entirely driven by generational differences among Republicans. Young Republicans are *far* more worried and engaged with the issue than older Republicans. Democrats though, across age bins—they're similarly engaged and similarly worried about it, and find similar levels of importance of the issue." In other words, younger people don't have the monopoly on climate concern, action, or even grief; but the stakes for us are very different than they are for our Elders.

Here we're concerned with developing our ability to communicate across generations: to build empathy and accountability, to share our stories, and deepen our relationships. We can practice in our own families—however we've defined them—with the goal of cultivating a kinship attitude toward all people younger and older than ourselves.

This is what adult responsibility looks like in the climate emergency.

Our Elders

Our Elders may be our parents or caregivers, or older nonparents, mentors, biological, foster, adoptive parents, or grandparents. Climate change was likely not a factor in their family composition because of *when* they came of age. Parents and grandparents in this group are watching catastrophes roll in on the future they had imagined for their children, grandchildren, or younger people who they love. They may be hearing from their own children that climate change is a factor in their family planning. Nonparents may have had other environmental or political considerations in choosing not to parent: the nuclear threat, for example, or the publication of *Population Bomb*, by Paul and Anne Ehrlich, in 1968 (we already let you know just what we think of this book, but it certainly was consequential for many people's decisions).

Being an Elder can be an honorific: at best, it means "having lived life, and ha[d] some experience and wisdom and hopefully have some skills to be able to share with the rest of the world to work on this problem," says Tina, an activist from Massachusetts. Elders Climate Action has chapters across the country; we spoke on a video call with a group from Massachusetts in early September. "There's a sense of solidarity with my generation that I take on, that I think I have some pride in," Roger Luckmann, a fellow activist and retired medical doctor, adds.

There is oft-cited data about retired people voting more often and being, on the whole, more civically engaged than younger US generations.[18] There are a few reasons for this, but a main factor seems to be time: those who have been able to retire usually have more time available to dedicate to organizing, more time for reflection, as well as more assets to protect. But what drives this generation, who won't live to see the worst of the climate impacts unfurl, to be involved at all?

Many Elders have gotten involved in climate organizing as a way to offer a positive legacy to the world, as well as to their own children. "What attracted me to [climate action] were two things," writes Rick Lent, who works with Elders Climate Action, "The traditional value of elders in a society, and the idea of the climate as our legacy. Today's elders are unintentionally leaving a legacy of a climate that will not enhance the lives of future generations—exactly the reverse. And it will not be like the one they inherited from their ancestors," Rick continues in a series of reflections that he shared with us. "Our generation can be said to have made a mess of things for future generations."

It can be tempting to let our Elders take all the blame. But they inherited the economy and ways of thinking that drove this crisis, just as we are inheriting it now. And many of us learned to tend nature, love justice, and question authority from those same Elders. How does each generation take the ownership that's theirs, but also recognize that this struggle is many generations long?

There is no way to do this without a wholehearted commitment to empathy, up and down the generational chain.

Climate communications experts also point to the idea of our legacy as an effective route for getting people involved. "There's some interesting messaging work on legacy framing. Like, 'how would you like to be remembered?'" says Dr. Goldberg. We are on a video call, and can see the whiteboard behind him in his campus office, with an upward graph scrawled in dry erase marker. "I think older folks are thinking more about their legacies than younger folks are. . . . And there is a strong push within the family to do the right thing on this issue."

The question of "the right thing" is something worth underscoring, we think. Climate change was well underway before today's Elders were adults, although it certainly radically intensified since the 1970s. Elders may arrive at a commitment to a climate legacy through a sense of responsibility, or obligation. Roger Luckmann reflects: "Climate change is the existential issue of our time," he says, "and I owed it to the world and myself and everybody I know around me to jump in and see what I could do."

Many resolve these feelings—or attempt to—through action. There's a sense among the Elders we talked to of practical, on-the-ground work needing to be done, of putting their shoulders to it and getting to work. Many of the people who are Elders now, at the time of writing, were also activists in the 1960s and 1970s, and describe a change in their own perspectives as organizers: moving from passionate late-night theorizing about Marxism to days of advocating for state climate legislation or doggedly door knocking for an issue campaign. People we talked to described themselves as having grown less ideological and more practical. That is neither "good" nor "bad"—it simply is. And it offers insight into how Elders may understand their own roles in the climate fight.

In addition to having amassed a lifetime of experience, some Elders credit climate organizing for helping them learn new skills, even in retirement. "I feel like I have dramatically increased my knowledge of the political process," Roger says, describing his work on a legislative advocacy committee, "and I feel much more able to really understand what's going on. And even more committed to digging deeper into policy development and ways that we can potentially influence our legislature to go in the right direction."

However, and despite the knowledge and rich experience many Elders possess, they may also be prone to forgetting—and by this we simply mean forgetting how one's age and generational position shapes one's perspective on what's going on and how much it matters. We are wired, as humans, to forget pain; we are wired to shake it off whenever we possibly can. When we asked Chris Begley about the challenges of climate relationships across generations, he reflected that our experience of major events—climate among them—is really shaped by what phase of life we're in when they happen. For Chris, for instance, the financial crisis of 2008 was a relatively mild bump: he had a job, had gotten tenure at a university. His retirement took a hit but bounced back. "And

I think partly that's where you get the boomer-millennial fights coming in," he says. "The relationship between what's happened and how it affects you is radically different if you're at a different point in your life."

Being any age requires empathy in all directions. Additionally, we have a responsibility when we are Elder to anyone—and that's all of us—to engage our youngers by listening first, and affirming their experiences. This doesn't mean pretending to agree with everything: it means putting aside preconceptions and agendas in order to make space and listen fully. Throughout the rest of the chapter we talk with experts about how to do this constructively.

Our Kids

Our kids are infants, children, or adolescents today, and those who have yet to be born, whose adult lives will mostly take place after the year 2030. We may be family or not, but we are responsible for them. The length and quality of their lives will be determined by the actions that we, as adults, take now.

Many of this cohort are *already* activists, like the plaintiffs who sued the federal government for climate harm in *U.S. vs. Juliana* or youth activists Vanessa Natake and Greta Thunberg, as well as children across the world who are on School Strike for Climate. They will make their reproductive decisions with a fuller knowledge of the human prospect—provided, of course, that they have access to education and family-planning services. For them there will be less uncertainty, but probably much greater harm. It is all of our responsibility to minimize this harm.

But a lot can happen before they are of age; the children in our world are vulnerable to many other kinds of climate harms, now. Psychologists and climate researchers agree: climate change is a threat to children's physical and emotional health, in ways that are qualitatively different from adults.[19]

Like most threats, climate threats are not distributed equita-bly among the children on the planet. Eighty-five percent of the world's children live in the Global South, where they are especially vulnerable; the World Health Organization estimates that children suffer more than 80 percent of the illness and mortality attributable to climate change.[20] These children experience multiple climate shocks, which are often combined with poor essential services like water, sanitation, and healthcare, a very dangerous combination.

Our children are in an impossible position: they are growing up with the knowledge that climate change poses an immediate threat, and they know that they will bear a disproportionate burden. That's a hard thing to process for an adult, let alone a very young human being. And yet, because they are children they have much less agency than adults do, and they are seeing the problem get worse, not better. And in the face of this unjust suffering, a lot of us adults don't know how to help, or even what to say.

Talking with children

Our kids don't need us to fix the crisis single-handedly to be good parents—or good adults—but they do need us to acknowledge it and help to fix it. The two tools we need to accomplish this are: *age-appropriate strategies for talking about the crisis*, and *modeling engaged behavior*.

Age-appropriate strategies

Dr. Susie Burke works in disaster psychology. She is a cheerful Australian woman in her mid-fifties with stylish glasses, who, after a career in psychology research, now runs her own private practice. We talked with her when it was late summer in the United States and winter in Australia; she was cuddling a hot-water bottle while we were in sleeveless shirts.

First, she says, parents need to be honest with their children, even the young ones. It's damaging, Dr. Burke says, to be "promising that everything will be fine. Or promising that there will be silver-bullet technological solutions that will come and rescue us, because we don't know and we can't wait for that. Because for young people, that makes them think you're talking bullshit."

Children are not dupes, she reminds us, and you can't shield them from this. "I can see how tempting it is for parents to do, and I hear parents doing that a lot. Because they're wanting to believe it themselves," she continues. "But it backfires. Because parents *need to be trustworthy.*"

Children need to be able to believe the adults in their lives, and so being lied to—whatever the intentions—makes them believe that nobody is taking the issue seriously. If children find adults untrustworthy, she says, "then they stop *listening* to their adults. And that's not helpful either. Because adults do have a lifetime of learning: about how to manage difficult feelings, and to build their resilience in stressful times, that children can benefit from. So it's not helpful if the children start dismissing the parents."

As Juan-Pablo pointed out earlier, a lot of us feel that our own parents fell short both in acknowledging and fixing the problem. Struggling to do this ourselves for our Kids can give us empathy for our Elders. And succeeding in doing this for our Kids can be a way to re-parent ourselves, healing by doing the work that our parents may have left undone.

Dr. Daisy Bassen, a Rhode Island–based children's psychiatrist with three children of her own, agrees that good information is key. When talking with her own children about climate change, she says, "our general approach has been: where are they developmentally in terms of how they understand the world? How do they see themselves in the world locally, and then in the larger spheres? What are they capable of, cognitively? And then: how do we present things?"

But telling the truth, as both experts openly acknowledge, can be scary, and it has to be done in a way that matches children's development. Dr. Bassen continues: "Our house is influenced by the fact that pretty much everyone is anxious to various degrees.

So: how do you talk about things in a way that doesn't create such an overwhelming sense of anxiety that it seems hopeless and paralyzing? Like, we would probably not spend a lot of time talking about 'part of the ice shelf fell off today!' because my oldest child would probably be able to rationalize it for a while, but there's a decent chance my youngest one would be in tears."

To deal with this, both Bassen and Burke suggest starting with connections that children already have. That could mean organizing within their schools for a farm-to-table lunch program; it could also mean putting on a tree-planting day. It could mean taking your child to your State House or City Hall in support of a climate proposal. It could mean bringing them along on an afternoon of door-knocking. One of Meghan's favorite political moments from her first term was when a troop of Girl Scouts came to the Senate Environment and Agriculture Committee hearing to testify in support of a balloon ban. The trick here is to help children feel activated—and included—in the world that's close to them.

False promises and reassurances that "everything will be okay" has another effect, even more damaging: it can reinforce children's belief that adults don't care, and that they are left alone in this crisis. That feeling of isolation, Dr. Burke emphasizes, can give children the impression that "things are really, really desperate and that I have to put my entire life on hold because I'm going to have to do it because nobody else is doing it!"

This is an important distinction, both for Dr. Burke and for many of the people we've spoken to over the years: can we take climate action from a place of love and solidarity, or will it always only come from a place of fear and desperation? "We really don't want children to be thinking, 'I have to do all of this because nobody else is doing it,'" Dr. Burke concludes. "We want them to think, 'I need to do this because it's the right thing to do, and because the planet is awesome and I love it, and I'm going to spend my life finding ways to be able to actively engage with all these other people that are doing it as well.'"

And if we want this for our kids, we have to create it for ourselves, too.

All of our children will live with a changing climate in their lives, for the rest of their lives. We can normalize dealing with it responsibly, as a persistent piece of what it means to be a human on the planet. The key here seems to be telling a complete story: acknowledging the frightening parts, the exciting parts, and the yet-unwritten parts can make the whole much easier to accept. In other words, it's easier to honor the wonderful parts of being alive when there is respect afforded to the bad parts, and vice versa.

Modeling engaged behavior

Both Burke and Bassen agree that the other most important thing parents and caregivers can do is to actively model, and include their children in, joining climate action outside the home.

"The best thing their parents can do is actually just to get really busy themselves with their own climate action—and to be doing *collective* action—because children need to know that their parents care for them, and also care about protecting their world," Dr. Burke says firmly. This has the dual benefit of both reassuring children that their grownups *really do care*, and teaching those children how to be engaged participants in building a better world.

Dr. Bassen agrees: "Don't necessarily get so bogged down on small, individual choices, because those won't necessarily have an impact! Figure out what groups you can join, or how you can support larger things, to really be effective. Which I think is both *harder* for little kids to understand from a systems issue, but in a way *easier* to understand. You say, 'We're joining this group, and this is what the group is doing. The group is trying to fix this problem, and this is how you could be part of that group.' I think they can understand the *group* part as opposed to understanding the big systems."

She playfully straightens her posture as she says, "I'm going to be a psychiatrist about this for a minute, but we need to talk about the dialectic between change and acceptance." She clarifies: how do we work on making progress, while accepting this is where we are right now? How do we acknowledge and meet the needs of those who are immediately affected, while also looking at the origins of the problem? As kids get older, their capacity for this understanding develops. "Maybe when kids are younger, you're focusing more on: how do we concretely help people who are already suffering? And maybe when kids are older, you start working on: how do we change the system so it prevents the suffering from going forward? Because a five-year-old can understand 'let's collect food for the food pantry because these people were in a storm,' and a fifteen-year-old can understand 'we've got to work on getting more wind energy, and so maybe you want to go to that protest because some people don't want a wind farm at the beach,' or something. I think it requires a lot of work for parents, to be honest."

As children get older, being part of climate work with their parents or caregivers can be very empowering. Some strategies:

- Tell the truth. Do so in an age-appropriate way, but tell the truth.

- Spend time on both *doing* (problem-solving, organizing, fixing) and *feeling* (curiosity, fear, grief, hope, excitement, problem-solving).

- Focus children on their own community, and talk about what that means.

- Think about immediate needs relative to long-term needs. Balance both. Be clear about them.

- Try different kinds of engagement for the right fit, maybe by picking from among the Beloved Community, the Big No, and the Big Yes. This approach might also help you feel more balanced in your own efforts.

The other thing that bringing kids into organizing can teach, Dr. Bassen points out, is humility. "Sometimes we think something's going to be effective, and then it doesn't turn out to be as effective as we want it to be. So then you have to go back and look at it again. I feel like we spend a lot of time telling kids 'it's okay to make mistakes!' but then we don't actually really act like it's okay to make mistakes. So I think it's helpful to really model that," she laughs.

Talking across Generations

In our conversations with activists over the years, a strong theme has emerged. One group is singing songs of innocence, while the other sings songs of experience: as activists get older, they tend toward less theory and more practice, or toward fewer structural interventions and more realpolitik, until eventually the radical kids and the jaded seniors can hardly recognize each other, much less organize together.

The experts have two clear suggestions for talking about the climate emergency with people of different generations, or whose experiences are different from our own.

First: start with shared experiences or shared values.

Second: listen.

Starting with shared experiences or values gives people a way to connect, even when the chasm between them seems vast. "Knowing what the other person cares about is where you need to start," says Dr. Goldberg. "For instance, knowing that the other person cares about God's creation or the sanctity of the earth, anything like that." Seek a meaningful point of agreement and connection to each other.

But Dr. Goldberg is quick to remind us that an entry point could also be purely pragmatic. "I remember talking to my neighbor about their solar panels," he says, "like, 'How do you like them?' They said, 'I don't really care. It was cheaper on my bill,' and that's it! Sometimes it's that straightforward. So you have all these different ways in, and I think knowing what your audience values and cares about is important."

Opening up yourself is an invitation for others to meet you halfway. "Sharing, and making yourself vulnerable, can actually be an effective strategy," says Dr. Goldberg, "because it's not really socially or morally acceptable to bash that person for their own experience, particularly if it's been difficult!" People who care about each other are more likely to be able to hear each other out, especially if they've created a rapport based on some shared value or experience.

This leads to the second approach.

Listening is the other key to connection. We asked Dr. Burke about how Elders, for instance, might respond to their child or grandchild who says they've decided not to have children because of climate change. She says that the first thing to do is simply acknowledge, without trying to justify, convince, or reframe, and without being defensive.

"We need to use our emotional intelligence," she says. "It's about being able to *validate* and *normalize* people's feelings." Young people are concerned for very good reasons, and it is incumbent upon all of us to recognize and to be "doing that work of emotional intelligence where we say, 'It makes sense that you would be feeling that!" It's about simply receiving and validating the other person's truth without trying to change it. What we *don't* want to encourage in these cross-generational talks is to pathologize climate grief and fear, and we don't want to sanitize it.

We are standing on the precipice. We still have pain to express, and a need for empathy. We have been facing climate change as an irrefutable fact of life since we were children, and one thing we've been hearing since we were children ourselves; "What a mess we've left for the youth to clean up . . ."

But! Anyone saying this hasn't left yet. Are we leaving soon? Let's get cleaning now, then.

We got a masterclass in intergenerational solidarity from Gayle Golden, a journalist from Next Avenue who wrote an article aimed at grandparents in 2017. A woman nearing grandparent age herself, she interviewed us, listened thoughtfully, and asked us curious questions. A few weeks later her piece was published, and we opened the link with some trepidation. But we were surprised and relieved at the gentle call-in that she was offering for other Elders.

"We need to be ready for our adult children's anger, for their anguish, for their questions, for their worry, for their grief and ours," she had written. "In other words, maybe it's time for all of us to *get our grandparent on*, with or without the baby." The article continues:

> We can also do more. We can help them with the messy tasks, which includes fighting to reduce our reliance on fossil fuels. And we can pass on the wisdom we've no doubt gained from decades on the planet: that life rarely hands us exactly what we planned; so when that happens, we make a new plan.[21]

This was a loving, moving response. And Golden also gave us the clue that we needed to get an early start on taking her advice ourselves. We are all links between the past and the future. Intergenerational labor never ends: our job is to accept, carry, and then eventually pass the baton in a really, really long relay race.

"One of the things that was really clear to me," says Rev. White-Hammond, "is that when God sends someone into your life to be family, you make them family. And so that has both allowed me—in some instances—to spread the level of privilege that I have to other folks who were not born into families with the same level of privilege. It has also allowed me to be in a relationship with children who I love like my own. Who have motivated me to go out and fight and do and be different and better."

To strengthen our links in the generational chain, this is the commitment we're called to make toward all our kids. Whether or not we share White-Hammond's faith in God, we have faith in each other. And it's never too late to repair the links that join us to our Elders. Josephine explains how her relationship with her father continues to deepen, years after his death:

> When I remember the love that underwrites my family relationships, I am more loving with myself. I see my own unbalanced books with new eyes. In one column is the insistent awareness that my life is too big and harmful, and in the other, the self I've built from old values: ownership, worldliness, creative expression, individuality. The happiest days of my childhood where when my father Enrico would come home from trips: from two months in the Afar desert in Ethiopia, with a beautiful doll for me and a doll-size string bed; or from Valdez, Alaska, having missed my 6th birthday, but bringing photographs of veterinarians using dish soap to shampoo crude oil off sea otters. It didn't matter to me til later that he'd been hired by Exxon for that job.
>
> He worked through the heyday and decline of human society's greatest profession—freelance photojournalism. His whole life was propelled by *Time* and *Life* subscribers' desire to see color photos of remote places and people living differently than themselves. He could offer greetings in probably twenty languages, though he couldn't always understand the replies. The message I received, by implicit and explicit means, was that the world is mine to see, that beauty is for acquisition, that travel and documentation have inherent value, that my perspective is broadly significant. I do and do not think this is true. I cannot make either column equal zero, or equal the other.

A friend read a draft of this essay Meghan and I have been working on, and he sent this response:

"I like [your] idea of [a] 'middle generation,' taught to comply and excel within the structure that's eating the planet. Why those values need to be rejected, and how hard that is to do—because you and I have to kill a part of who we are, and who we've learned to be, to do it. It hurts to think about this. It should."

It's elegantly stated, and I know that the world could never support another generation like us or our parents, but I resist the call to kill that part of myself. It's not a horcrux. Even if I were able to kill it—and I'm not sure how that's done—I don't know how I could make myself want to. Because it's my father. Because that imperfect part of me that is him is also the part that loves the world enough to soldier on.

One time when I stopped at Mt. Sinai on my way to the airport because I couldn't extend my visit any longer and he still hadn't been released after another respiratory infection, he told me, "I've always admired the scenes you draw. You frame them from such unique perspectives." At the end of that visit, as I waved from the doorway, he held up his hand, searching for just the right Americanism, before shouting, "Keep fighting, baby!"

Just like I don't want to kill a part of myself, I don't want my life's work to be fighting. But it makes a good personal slogan when paired with something else he said to me at the beginning of the relationship that is now my marriage: *che se son rose, fioriranno*. Keep fighting, baby! That if they're roses, they'll bloom. I grieved him with my family, and then there was a shift. The end of grieving didn't mean the end of sadness or the disappearance of loss. But it did mean that I dropped a burden I'd been carrying: what remained was a living, changing relationship with the part of him that is me.

Now if I can love him, draw strength and meaning from that love, and at the same time see the harm he and his generation did by behaving as if the world has no limits, then I can use the same compassionate touch on myself. It's not possible to quit the human project, or hide from our responsibility as beneficiaries of it. It's only possible to reach up from our positions within it and try to pull the best of it up with us.

It is also true that no matter what grim work I must do, and whatever banal, deadening language is commonly associated with that work, my body will always move toward warmth, absurdity, trust, and encouragement. The elders who tell me the truth as they see it, and who affirm and encourage me to tell my own. The kids who laugh at my jokes and listen to my stories, then make fun of me for what I've never heard of. I grow toward the relationships that invite me to grow, that welcome what I have to give.

CHAPTER 6

See Yourself in the Beloved Community

I have never been afraid of taking action. I know that that's an experience, it's just not an experience I've ever had. And the other experience I've never had is not living in a robust community.

—*REV. MARIAMA WHITE-HAMMOND*

Climate Justice isn't about saving the planet. It's about saving humans.

—*JAYEESHA DUTTA*

THERE IS WORK WE EACH DO ALONE TO UNDERSTAND OUR PLACE in a changing world (chapter 4). But sit alone too long in the gathering darkness, and despair creeps in. *No one around me seems to be freaking out: am I the only one? Nothing I try seems to make a difference: what can I do?* Almost as daunting as the crisis itself is the hedge maze of isolation that grows around it. How lost we feel when we turn inward. How overwhelmed when we search for answers without knowing what to ask. When we define our experience in terms of the problem, the problem will take control.

185

It will overshadow us, deplete our resources, and sour our attitudes, defeating us whether we solve it or not. And so to find our way, the first step we each take is toward one another. By joining with people we already love and depend on, we address problems as we confront them *at our own scale*, using the skills we develop together, with the full force of ourselves, combined.

Many of us had at least a glimpse of this potential in 2020, when mutual aid groups sprung up across the world in response to COVID-19. People with no prior organizing experience were meeting their neighbors online to coordinate food collection, homemade mask drives, and schedules for volunteer helplines. Many went door to door, finding out what their neighbors needed and what they had to offer. Some of these networks are still active. When you are ready to take climate action, we invite you to take the first step: *get together*. Before we break the monumental job before us into Big Nos and Big Yeses, let's look at how we can find each other, come together, and stay together.

Every successful social movement in history has shared this same foundation: community. Josiah Royce, an American philosopher-theologian, coined the term "Beloved Community" in the early twentieth century, although it's Ella Baker, founder of the Student Non-Violent Coordinating Committee (SNCC), who popularized the idea. The Beloved Community's core tenets are economic and social fairness—twin pillars that support a just movement. Here, our own well-being is innately bound to the well-being of others. And in this sense, community is both a means and an end.

Baker understood this to be achievable when enough people—committed to relationships and trained in nonviolence—work for each other and the integrity of a shared vision. The Beloved Community abhors poverty, hunger, and discrimination equally anywhere in the world. It reconciles disputes peacefully through dialogue, rather than violence and militarism. The Beloved Community cherishes the equality, dignity, and potential of all. Some people call this social justice. Others call it the Kingdom of God. By whatever name, it's a vision of a world built by the love and care we have for each other and the earth that sustains us.

But as Baker and every organizer before or since knows, communities are full of humans, and humans are imperfect. No association of people will be wholly free of conflict, tension, or strife—nor can we expect it to be. Conflict is our opportunity to put nonviolence into action, to clarify our views and assumptions, and to transform opponents into allies. Through this practice, we can make ourselves and our relationships stronger.

This kind of collective accountability gives us the strength and the inspiration to act, and the numbers to make action effective. Hope—the belief that what we do matters even though we can't fully know the outcome—is a natural by-product of such a community.

WHY COMMUNITY MATTERS

To be human is to know and love other people. We make lives together, and we give each other meaning; we cannot be healthy alone. We experience our own belonging and purpose through these personal ties, but the effects of belonging and purpose are actually visible on the collective scale, too. This does not mean that all relationships are good, or that groups are always safe, right, or fair—we all know that they can exclude and discriminate. Communities need to be bound by laws, rules, and values that ensure fairness. But it *does* mean that the future we create arises from our collective vision, so that vision must tend to our relationships.

Social capital is one of many names we give to the quality of our connections with each other. Research shows that the single most important factor in a community's health (by measures from educational achievement to local economic strength) is the level of social capital within that place.[1] In an influential study, political scientist Robert Putnam showed that many factors traditionally used to explain success—location, history, leadership, enrichment programs, economic advantage—made only a marginal difference in the health of a community. Rather, a community's well-being depends almost entirely on the quality of relationships among its members.

Flexibility and trust

Communities—whatever initially brought them together—are tremendously flexible entities that can do all sorts of work. The successes of the US Civil Rights Movement, for example, were built on the bedrock of the Black church: people originally came together to worship, but through fellowship, were able to adapt church-born practices and relationships to meet the needs of the moment, creating carpools, potlucks, and phone trees that sustained and propelled the movement, and the Montgomery Bus Boycott specifically.[2]

Despite the loss of civic space and spirit over the past fifty years that Putnam famously described in his book *Bowling Alone*, we are surrounded by contemporary examples of triumph through fellowship. When Jay O'Hara, a New England activist who we met earlier, faced trial for his maritime act of civil disobedience, he drew strength from his contra dance friends. "Since 2001 I had been on-and-off engaged in a New England folk tradition called contra dancing," Jay explains. With his sharp gaze, long brown ponytail, and rail-thin frame, it's easy to imagine him performing an antiquated dance in a crowded hall to lively folk music. "Over the years, some [of us] started hanging out outside of dancing and grew into a genuine community, spread across the northeast US, caring for one another, and sharing a music, dance and song culture that bound us together."

In 2013 when Jay and Ken Ward dropped anchor in the path of the coal-carrying *Energy Enterprise*, their little white lobster boat blocked the ship from docking for almost a day before they were apprehended. After their arrest, Jay and Ken spent months preparing for what they hoped would be a precedent-setting climate trial. They were planning to use the Necessity Defense—a legal strategy in which they would argue that yes, they did block the ship illegally, but that the unchecked climate crisis and lack of political action left them with no alternative but civil disobedience. They never got to use the Necessity Defense, though, because the prosecutor dropped their felony and other criminal charges in a surprising act of public solidarity.

Earlier we discussed the big team of activists that made this action possible, but for their day in court, Jay says, "the contra dance community showed up in force. They prepared food for those who were in town, joined us in the courtroom, held us in their hearts as we negotiated with the prosecutor." And when Jay and Ken walked free and their press conference was all over, Jay and his fellow dancers gathered in a circle on the courthouse steps and led a stirring song from their shared repertoire:

> *Oh my dear friends*
> *I truly love to hear your voices*
> *Lifted up in radiant song*
> *Though through the years*
> *We all have made our separate choices*
> *We've ended up where we belong*
> *Walk with me and we will see the mystery revealed*
> *When one day we wend our way out to the farthest field* [3]

It is unlikely that those dancers expected their hobby to lead them into court for a climate trial. But strong communities can do this: they flex and reach to meet the needs of their people.

Accountability and risk

Available evidence tells us that the main reasons we'll take risks in pursuit of social justice are our personal ties. Human beings only make such sacrifices when we love each other deeply. Sociologist Doug McAdam and colleagues found that people who joined the

1964 Freedom Summer—and who stuck with it—did so because they had close friends who were also involved in that work.[4] This dynamic—of people taking risks for each other—shows up across the world and in every kind of group, ranging from 1970s solidarity networks in Switzerland to the Paris commune insurgency in 1871.[5,6] Whatever our aims, we're much more willing to take bold action—indeed any action—when we do it with, and for, people we care about.

Communities also strengthen accountability for behavior. The Reverend Dr. Chris Davies is the Executive Minister for Programs and Initiatives with the Southern New England Conference of the UCC. She told us about one youth group with whom she had worked, noting that the community itself was designed to elicit committed, healthy behaviors from its members. Young people were able to overcome alienation and loneliness and to call forth their better selves. In that kind of space, the conversation was focused on connecting across differences, creating, Rev. Dr. Davies says, "space for everyone and beyond." Those youth weren't necessarily focused on faith; the Reverend sees her primary job as creating a container in which people can explore themselves in relation to each other. "The container helps," she says, "especially for young people. Because within that, we can be clear: All people are welcome, all behaviors are not."

These practices are not restricted to religious communities: they can exist within any space, literally. In a qualitative study of teenagers in five major cities, Richard Flory and Donald Miller found that millennials are not "the spiritual consumers of their parents' generation, rather they are seeking both a deep spiritual experience and a community experience, each of which provides them with meaning in their lives, and is meaningless without the other.[7] In other words, when we say we are not looking for a "faith community," millennials and Gen Z might mean we are not interested in belonging to an institution with religious creed as the threshold. However, we are decidedly looking for *spirituality and community in combination,* and feel we can't lead a meaningful life without it. If religion isn't your particular brand of vodka, never fear: this kind of community can show up everywhere from the pickleball court to the barbershop to your Parent-Teacher Association.[8]

Isolation, alienation

The flip side to the grand benefits of community is that the *loss* of community, particularly in industrialized countries, is a key cause of social injustice, alienation, and harm. We lose our social ties for lots of reasons, many specific to our time: we work longer hours for less money and have to hustle to survive; we're priced out of our towns and neighborhoods and wind up far from each other and isolated by bad transit; we rely increasingly on media and telecommunications rather than meeting face-to-face, and we're mutually enraged by our politics, to name a few factors. These pressures intensify our isolation and exacerbate the self-focus of our institutions and occupations. Put simply, we get sad without connection.

Loneliness and sadness don't help us solve problems, and they certainly don't equip us for change-making. When we don't feel belonging or commitment to each other, we are less able to do courageous and creative things. It's when we belong to each other that we take risks and generate ideas that matter. Not only does loneliness make us miserable, but in isolation we deprive each other of our gifts. Talents atrophy when we don't use them, when and where they're needed.

Our most urgent task is to step from isolation to connection.

Quality over quantity

Now is a good time to note that if you're an introvert, this doesn't have to be your personal nightmare. We are talking about the *nature of our relationships,* not the frequency of our get-togethers, or the number of people invited. Community doesn't mean surrounding ourselves with thirty acquaintances at all times. Josephine is often introverted: she loves solo time and one-on-one conversations and gets easily overwhelmed in groups. When our work involves speaking in public or corralling a crowd, sometimes she challenges herself to do it, and sometimes she relies on Meghan, a natural extrovert, to take the lead. But running Conceivable Future involves a thousand kinds of jobs, public and private. Because we know and trust each other, we can be flexible in a way that challenges and supports us both.

Lex Rofeberg is a digital educator and rabbi who cofounded and coproduces the podcast *Judaism Unbound*. He has a lot to say about this. "I'm obsessed with the word *community*," he tells us. "I'm obsessed with the ways it's misused or overused—It's a nonsense word. Jewish 'community'—when people say 'my Jewish community,' they tend to mean their synagogue." Lex is a warm, exacting, quick-speaking man of thirty-two, who swears easily in conversation. He has an intense gaze and wide smile, which tempers the sharpness of his next words. "I hate it! Because what it does is equate the word *community* with just a bunch of people who gather in a room regularly. And that's it!" He continues:

> I want community to *actually* mean a group of people that not only exists and shares some characteristic, but rather is there in times of sadness, shares times of great joy. And in the middle time! So often the emphasis is on: "do you have people to lean on when things suck? Do you have people to share your birthday party with, and other happy moments?" But I actually think that sometimes it's about just having people for the middle time.

In other words, a community is bonded by shared commitment to each other—in the grief, in the celebrations, and, maybe especially, in the boring daily stuff. In the carpool and the dog walks and check-in phone calls, the loaner tools, and keeping up with each others' families, work, and romance.

A community is greater than our individual experiences. It shapes and is shaped by us. The unified body to which we belong has power, even when we screw up and fall short of ideals. In community we can acknowledge our shortcomings with dignity, learn from them, make amends, and move on. It's important to acknowledge that resources are not fairly distributed between communities, and a lack of money, time, or space can make everything harder. But we find our single most important resource in each other.

As a grounded, committed individual is the basic unit of a community, so a strong community is the necessary building block for both the Big No and the Big Yes—we don't have political power to resist bad things or embrace good things if we go it alone. Walking together, we can remember that we are powerful.

STEP 1: FINDING EACH OTHER
That all sounds great, but many of us are surrounded by people and still feel climate despair. Because communities require accountability, they are also good at enforcing norms. Right now, the cultural norms around climate change are silence, paralysis, and inaction. You may feel alienated because your group is ignoring the flaming crisis on our doorstep. Or if you do belong to a climate organization already, you may not yet find a sense of belonging or fellowship. How do we transform a bunch of people into a powerful, change-making community?

First we have to find our people. At its most basic, a community is a group of people that shares *something*—whether that something is a neighborhood, an affinity for punk music or dogs, a taste for home brew, or a faith practice. Any existing association, from the Democratic Socialists of America to a snowboarding club, is a great place to start.

We'll use ourselves by way of example. Meghan and Josephine met through our mutual friend Jamie; it was a social, not an activist, connection. As we grew closer and our project developed, we drew in people from all aspects of our lives: other organizers and activists, friends, family, neighbors, former classmates, Meghan's bandmates, Josephine's yoga pals. At the same time, some of our closest friends and family avoid this project, whether because of the climate or the kid topic, which is to be expected. Any functional group is at least a bit self-selecting, whether by geography, capacity, identity, or interests.

Together we represent the center of a community we created, and we have collaborative relationships with other climate organizations across the country. But individually, we also occupy positions in many other overlapping circles. We are family members and family friends. We are alumnae of high schools and universities, and residents of separate neighborhoods, cities, and states. We each teach students in several different settings. We have colleagues, mentors, and mentees in our professional fields. We have school friends, work friends, childhood friends, and casual acquaintances. Meghan is a state senator. Josephine and her husband live in a building with twelve other households, in a politically active ward. Meghan plays trumpet in a twenty-five-piece activist marching band. Josephine meets with a writing group and quilts with friends. Meghan is Jewish and Josephine has attended the odd Quaker meeting. We both go to gyms and coffee shops, restaurants, and bars. We each have social media networks and mailing lists to which we subscribe. We belong to the groups of people who ride Amtrak, public transit, cars, and bicycles, as well as the occasional airplane. You'll notice that these are very different kinds of connections. But there are a few types of affiliation, and identifying them may allow you to see the power of your own relationships in a new way. Groups can be goal-based or process-based, hierarchical or democratic (or anarchic!), formal or informal. They can be tiny or enormous, share values or simply share fun. We describe types of relationships here, with a note about possible climate entry points. We do this to give you ideas, and also to point out that climate action is present in *all relationships* if we allow it to be. We hope this will get you started identifying the communities to which you belong.

We want a widespread cultural transformation, from feeling isolated to feeling connected and capable of doing bold things. This isn't accomplished overnight. And the climate entry points that we suggest here are supposed to be just that: entry points. They are ways to *start* thinking about who we are and what we're capable of. If our examples don't resonate, you know better than anyone what might work for you. These are not the sum total of our abilities, or a complete description of the work that needs doing (for a more comprehensive list, turn to the next chapters). But to connect and build our community strength, we start by warming up our imaginations.

Family

As we discussed in chapter 5, these are our deepest love relationships, whether that's our birth parents and grandparents, our spouses, children, siblings and niblings, or the kin we have acquired along the way. One fifty-one-year old lesbian woman we spoke with had been disowned by her parents in her youth because of her sexual orientation, and had created a network of beloveds who saw her through that painful time and have been her family ever since. She goes to them when she needs support, or when she needs a dogsitter. To their children she is an aunt, sharing Sunday dinners and Christmas gifts. This family danced at her wedding and cried with her during her divorce. One of the special strengths of family networks is that they are often multigenerational, while many other communities are not. Through our family we share the experiences of people who have lived—or who will live—in a different time.

Climate change is felt most personally through family relationships: we'll fight like hell for the health and safety of our loved ones, and against the burning, flooding, and paving of our shared places. Although some families volunteer, organize, or march

195

together, we view this as the group that helps us understand the stakes of the crisis and make sense of our experiences, rather than a necessary hub of action. How do we make our Elders proud and keep our children safe? How do we make sure our babies know what songbirds are, not what they *were*?

Social groups
Social groups are typically informal, and based on intimacy and affection. These bonds may be more casual than family: we might share feelings, rather than responsibilities. These are the people we consider *friends*—folks who we care about, who we see regularly, and to whom some combination of shared interests, values, or history connects us. Maybe we've known each other since grade school. Maybe we got tight during a grueling orientation session at work, or are bound by memories of questionable choices made during college. In any case, friendship supports trust and loyalty.

Behavioral changes only begin to matter on a collective scale, and our behavior is only shifted positively by people we trust. How can we work together to change our group's consumption and emissions? Shared household, transportation, childcare duties, or solar lease? A community tool library or buy-nothing group?

Religious and spiritual communities
Religious, spiritual, and faith groups have historically been the major source of community in the United States. Congregations are built on shared spiritual commitment, but can equally nourish social and emotional life. "Congregational community is in a space of evolution," says Reverend Dr. Davies. She is a young, dynamic conversationalist with a cheery, purposeful tone and bright lipstick. She lives in Cleveland, Ohio, which is also the national hub of the UCCs, and she describes herself as "queer femme, an urban beekeeper, and a creative networker of communities." As we search for ways to fill family and cultural vacancies, Rev. Dr. Davies sees her church as able to grow to meet some

of these new needs: "It provides grandparents when your own grandparents are far away, and extra aunties and uncles. Lots of compassionate love heaped upon kids especially, from people who are there to support."

Climate action is supported by scriptures of all faiths. Religious groups speak with unique moral authority, and can often also nourish intergenerational relationships that help us take better care of each other. Many congregations already do service and advocacy work: is climate justice a priority in yours? If not, can you start conversations to bring it to the fore?

Classmates and coworkers

Classmates and coworkers: most of us have at least a few in either category. These relationships can sometimes be formal and hierarchical (most of us have bosses, after all), but they are unique in that we often spend the lion's share of every day with our colleagues, rather than our family or even our friends. Some of these environments are large, diverse, and/or intergenerational, others are small and/or homogenous.

Lex remembers how his mother, an employee at SC Johnson, built a community among her few Jewish coworkers. "They would find my mom somehow!" he says, laughing. And his mom was not observant. Rather, she came from a family that had left the Orthodox practice to become socialists—but still hosted dozens of people at the holidays. His mother continued this tradition by inviting her Jewish colleagues home. "To me, that's just obviously something you do," Lex says, "but it's become clearer to me that a lot of people in the world—they don't feel that they owe anything to newcomers to the town. It's a much more private [idea of] family, your family. And if somebody's not in your family, they don't have a place."

Lex's mom used a religious practice—like a Passover seder—to extend welcome to office newcomers and ultimately built a community tradition from her house. And although this is a religious example, there are countless other affinities that can bring people to the dinner table.

Questions to discuss with your colleagues or classmates: Does your school, workplace, or labor union have a climate adaptation policy and a disaster preparedness plan? Does it have sustainability commitments, a fossil-fuel-free endowment, or retirement accounts? Does it participate in a community-based renewables program?

Support groups

Support groups can provide needed connection during painful or life-defining struggles, such as a death, trauma, or an addiction. Relationships in these communities, even across class, race, and age lines, can be particularly strong because they require vulnerability and consistency from the outset. Recovery organizations particularly foster very strong personal ties, through practices like sponsorship. If a group has been together for a long time, it may choose new topics to explore, but a support group is effective primarily because of its deep focus on a specific shared experience.

Climate grief and anxiety can complicate personal trauma. Acknowledge this, and see how it deepens your group work. You may discover that the skills your group has already mastered can be applied to the emotional dimensions of climate change. Also consider: would it feel good to have a climate support group?

Neighbors

Neighbors can form special bonds and can also present special difficulties. Like a school or office, a neighborhood can bring people with different beliefs and identities into regular proximity. If you live like Meghan grew up, your closest neighbor may be a mile away, or if you live like Josephine grew up, your neighbors could be fourteen floors of people in one apartment building. Whether you can hear their toilet flushing or barely see their mailbox, these relationships need special consideration. We are often more reserved with neighbors than other acquaintances: if we fall out we still have to see each other regularly, forever! On the other hand, we care for a piece of the earth together.

Americans are much less rooted than we used to be—unlike our grandparents, millennials are likely to move more than ten times in our lives—but a sense of place can be a strong bond with the people who share it. Do you feed a neighbor's cat when they are away, sign for their packages, or plow their driveway? Do you join barbecues or block parties on your street? Who do your kids catch the school bus with? Who comes trick-or-treating at your house? Your neighbors share part of your fate. Climate change links us most immediately to neighbors: we share air and water quality, infrastructure, flood risk, and the price of homeowner's or renter's insurance. Even if we disagree about an election or a property line, risk mitigation and environmental improvements are common sense at the local level. And everything we do to generate goodwill among our neighbors makes those tasks easier.

Neighbors are often first responders in emergencies. What might they need, and what might you need, if the power was to go out for a week? Who has small children, or special medical needs? Getting to know your neighbors is one of the first and biggest steps toward community resilience.

Political and civic organizations

These groups are often campaign-based (get this person elected, stop that bad piece of legislation), but can create a particularly powerful kind of connection, because sharing a political goal (the campaign) requires you to share values, responsibilities, and a vision. Political commitment can foster deep friendships as well. Jennifer Boylan, an energetic mother of two young children in southern Rhode Island—and a newly elected state representative—tells us that joining a political organization changed her life in ways she didn't expect. "A huge part of it is feeling like I found my tribe," she says. Jennifer came to activism through Moms

Demand Action for Gun Sense in America in the aftermath of the Sandy Hook shooting. Through that group she expanded her political and social connections. "These are my people" she tells us. "They feel the way I do! And I've made so many amazing friends, and they are all made of the same stuff, and we all work together, and we work hard. We have fun, too! . . . There's an element of camaraderie that's fantastic, as we're doing this good, powerful, meaningful work. It's very satisfying."

Every political and civic issue is affected in some way by climate change. If your group isn't already focused on climate, think through what it means for the issue you care about, together. Adapt your priorities, or even mission accordingly. This is *not* a call to drop what you're doing to do something else. Instead, if you fully acknowledge how climate change affects your work, you may find new allies, a stronger analysis, and renewed commitment.

Shared experience
Shared experience can create a shared *identity*. This could mean alumni or veterans, for example, or people with a common experience of social exclusion. Meghan went to Smith College, where a treasured on-campus ritual was Friday Afternoon Tea with classmates and roommates; many alums across the country hold weekend teas in their homes, using those meetings as a chance to network, learn about local legislation, discuss books, or just be in company. Women politicians often connect via regional networks to find or become mentors, talk shop and start new projects, and bond through their shared experiences.

A shared experience or identity may offer your group special skills for climate action, or insights into a particular aspect of the crisis. For instance: wheelchair users may have unique knowledge of the holes in their city's public transit and sidewalk systems. Many veterans understand the connection between oil dependence and militarism better than civilians. How does your shared experience give you insight into what's happening, and better empower you to act?

Shared interests

Shared interests bring people together easily, often because they involve fun. This could take the form of a creative collaboration project, classes, or a professional association, for example. Do you belong to a dojo? Do you fish or ski? Do you LARP or play dominos? Every hobby is better enjoyed within a stable climate, and can be a starting point for climate action.

One example we like: in Meghan's city, one woman named Meara began a running club with other dog owners in her circle of thirtysomething neighbors. They would meet weekly with their dogs, and go running in the park. When a local brewery opened, they began to go for post-run beers (also with their dogs). Now they describe themselves as "a run club with a dog problem." Their mission is "to run, collect donations to shelters and animal rescue leagues, and then to drink or socialize and build a community." Whether the mission is dogs or climate, people brought together by shared interests can become a solid foundation for bigger social action.

What skills and resources does your group possess that might make the experience of changing the world more fun? Many martial arts practitioners are skilled in de-escalation, which comes in handy everywhere from meetings to marches. Your moms group may want to knock on doors for a political candidate that inspires you. Your bicycling friends could lead a group ride through the neighborhood to map heat islands, or places where the bike lanes need improvement.

You may also share an interest with others that you usually practice alone. That is great: solo skills are always needed in a group effort. You can elevate the aesthetics of any community event with handmade swag or silk-screened posters. Every anti-fracking campaign could use a better website, if you know how to build one. Share your beauty with the group through writing, crafting, data visualization, or cooking.

Online spaces

Social media is where far-flung people with shared values can learn, organize, and enjoy the world together. Online spaces can be important nodes of connection, creativity, and forward-thinkingness. When Lex's team launched a podcast, he says, "it 'succeeded,' in that we reached a lot of people pretty quickly—we had enough listeners that people started throwing around terms like [the] 'Judaism Unbound community.'" But he wasn't sure they'd arrived at community status yet. "I know what they mean," he continues: "that there are people who listen to our podcast and if they were to meet each other they'd have an instant thing to talk about. It's like when I meet somebody who roots for the same sports team I root for. Like, when I'm traveling, I wear stuff either Milwaukee-related or Providence-related, something that shows a thing about me."

But the "online space of shared interest" became "community" with repeated time, exposure, and empowerment. Lex had always believed that digital spaces can be communities in the abstract, but the podcast showed him how it happened, in real time. "I thought of our podcast as an educational mechanism more than a community-building mechanism for the first four years," he says. "But I eventually realized that while people were learning things they might not have known—that really they were learning a new [way] to think about Judaism, and that they themselves were empowered to create it. And then," he says "people were knowing each other a little better and making good mischief in Jewish spaces, both offline and on. And that is really powerful. It got me to start thinking of us as a community."

A great, secular example of a shared interest group becoming a shared action group is the online community AF Gang: formed as a fan club for the British rock band IDLES, the now 30,000+

member international group is famous for its pro-mental health, anti-stigma, and antidiscrimination activities. "In spite of it all, life is beautiful," they say.

Digital spaces have long been havens for groups that are either geographically or socially isolated—everyone from gamers to members of the LGBTQ+ community to people with disabilities. An inclusive understanding of community recognizes the beauty of digital connections.

Where do you most enjoy your time online? Who in your digital networks shares an affinity or values with you? Ways to bring your climate awareness into your online life abound, whether through friendly chats, promoting campaigns, or following Mary Annaïse Heglar's lead and "greentrolling" Big Oil on social media.[9]

Transactional relationships

Finally, we include transactional relationships such as customer, client, member, and subscriber, even though they can be almost the spiritual opposite of a true community. We do so for two reasons. First, there's a big overlap between these and other modes of interaction: we're likely to have transactional relationships with family and neighbors as well as strangers, and it's important to recognize this. Second, we often ignore our potential power in these systems: many of us just focus on our own "ethical" purchases. But every company has an aggregate perception of what their "valued customer" wants, which often guides its policy, and so we need to go public with our preferences. Many companies will move mountains to avoid or appease a public complaint.

For example, one Rhode Island neighborhood pressured its local convenience store to remove an over-the-counter sex "enhancement" pill because of its violent and sexist packaging. It took only about seven or eight complaints for the product to be sent back to the distributor for good.

In another instance, amid the protests that broke out nation-wide following Donald Trump's 2017 immigration ban, the New York Taxi Workers Alliance announced they would not be driving for one hour at JFK out of solidarity with the protesters. "Drivers stand in solidarity with refugees coming to America in search of peace and safety and with those who are simply trying to return to their homes here in America after traveling abroad," the alliance wrote on its official Facebook page. "We stand in solidarity with all of our peace-loving neighbors against this inhumane, cruel, and unconstitutional act of pure bigotry."[10] (This was also an excellent example of coworker and online action.) Uber, on the other hand, eliminated surge pricing at JFK, which many viewed as a way to capitalize on the taxi strike and perhaps even break it. The #DeleteUber hashtag circulated widely at that time as users deleted their accounts globally, forcing concessions from the company.

Publicly boycott bad products, talk to others about why, and show support for good local businesses. Engaging publicly is key: An individual boycott you don't talk about is called a preference. Mobilize community opinion (useful tools include: neighborhood listservs, letters to the editor, hashtags), and you could score an outsize win, while building people power.

STEP 2: COMING TOGETHER

What we hope to show here is how many positions each of us already hold—even before we start to plot a specific course of climate action. Take a minute, if you're curious, and make a few lists. Conscript a friend to do it with you if you want:

- What identities define me?
- What are my major relationships?
- What groups am I *already* a part of?
- What institutions do I have a relationship with?

- Where do I go regularly? How do I get there?
- Who do I communicate with regularly, whether in person or remotely?
- Who produces what I consume/who consumes what I produce?

Once you see the breadth of your social ties, you can start considering their qualities with these questions (Give the same answer repeatedly? That person must be beloved to you!):

- Who do I feel comfortable/safe with?
- Who helps me feel connected?
- Who do I have fun with?
- What am I in charge of?
- Who depends on me?
- Who would I ask to water my plants if I went out of town?
- Who would I call in a crisis, or who would I drop everything for if they called in a crisis?
- Who shares my values?
- Who do I influence, and who influences me?
- Who shares my skills? Whose skills complement mine?
- Who do I look to for inspiration, whether I know that person or not?

As you answer these questions and notice what emotions you feel, please recognize what you're *already* doing. The effort it takes to maintain relationships—to offer and ask for support, to share the good and bad—is rarely counted in our conception of "work," or "activism." But *nothing could happen without it.* Acknowledge where you *don't* feel supported or bonded, where you don't feel comfortable, and where you're not having fun. Odds are, you're not the only one who feels this way. What would need to change to create more connection?

Based on the work you did above, do you see a good starting place for yourself? How will you help a group you belong to grow toward a Beloved Community, ready for climate action? Or, if you don't see a path forward with a group you already belong to, can you use these insights to seek a group that fits better, or bring people together in a new way?

We emphasize that, especially for the beginning organizer, *it can be enough just to come together* and recognize our connection. This can happen with a new group, or with an existing one. A first meeting doesn't need to end with a ten-step plan to contain the entire climate crisis. Rather, it is an act of faith that getting together will help us understand what we need, and how we can meet that need.

Moving from "Group" to "Community"

The next step, now, is to transform from a *group* to a *community*. The difference, as Lex pointed out earlier, is that, whereas a group may just be a loose collection of familiar people, a community has accountability among its members, and to the whole. A community shares a sense of belonging as well as a vision—but a sense of belonging comes from recognizing people's whole selves, rather than just a point of affiliation or people's utility toward a common cause. (Remember Jay's dancing community showing up at his trial for civil disobedience!) Burdens are easier to bear when we share pleasure and support.

In our view, the most effective starting point is within an already existing group; deepening those relationships with a commitment to climate justice. In other words, the goal is to bring the conversation *to where you and your people already are*, instead of striking out alone. Go there first. Reach out and see if folks are willing to have a dedicated conversation about climate.

"We know not everyone is an organizer," says longtime organizer Jen Mendoza. "So how do we get people together? Every movement—that has ever been!—was started with people having conversations! And so how do those people make it happen? There's so many books that will show you the way. And we have access to the internet, and there's also tons of resources and guidance and information there. There's so, so much that people can dive into and figure out."

But not everyone is ready for a climate conversation, for a variety of reasons. Sometimes people's energy is all used up just putting food on the table for their kids. Sometimes other traumas make this subject impossibly difficult. Sometimes people just won't engage. If you want to be around more climate-focused people, then seek out existing groups who are doing the work you want to do. Revisit Step 1 to consider your different community attachments, and see if another group is more ready to embrace climate action. Because climate issues affect virtually every aspect of our lives, the range of possibilities for involvement are tremendous.

STEP 3: STAYING TOGETHER

Once we get together, the challenge becomes: *how do we become greater than the dangers we face?* Here we nurture these relationships so that they sustain us when things get rough.

If a community exists it *already* shares values, at least implicitly. Your Italian conversation group values effort and consistency; your AA meeting values candor and privacy; your professional association values continued learning and growth. And, as we said before, groups can also unwittingly enforce conventions that restrict us. When we explicitly decide what we value as a collective, we also get to decide what we do *not* value: which social norms are limiting our relationships or our imaginations. We can choose to put those behaviors aside. By articulating a group's values, you may find yourself in a smaller, closer, subgroup that pursues its goals within—or apart from—the whole. Articulating our values together—what we choose to *embrace*, and what we choose to discard—is the next step toward readiness for the fight of our lifetime.

The strategies we offer here, like everything else in this chapter, are suggestions and starting points for your own exploration. We've drawn them from community organizing experience and existing scholarship, but they may not be appropriate for everyone all the time. They are not exhaustive. But we have found them helpful for moving us from fear, stuckness, and disengagement, to more lively and radical connections.

Have fun

One of the classic pitfalls of activism is too much *work* and too little festivity. When work is a grind we disconnect emotionally, and look elsewhere for our "real" life. Strong communities and satisfying work depend on play and fun. Every birthday, holiday and victory can be a shared opportunity to dance, eat, sing, decorate. Bring the things that make you happy into your climate action life. Mark solemn occasions as well: your whole experience is welcome here.

Look for the gifts

Begin with yourself; offer what you're good at. Then, what is your *community* good at? What makes it special and unique? How can you help draw out those gifts in yourself and others? Sometimes called the "asset model" in community development, this perspective emphasizes people's talents and skills, rather than our deficiencies or shortcomings. We take a deficiency-based view in many of our most powerful institutions: medicine "fights" disease rather than prioritizing prevention, prisons punish people instead of rehabilitating them. Focusing on what's "wrong" with us induces shame, and we know shame is paralyzing. We build strength with the *assets* of our people: our gifts, skills, and knowledge, and what possibilities those things open.

Leadership is convening, not dictating

A culture of retribution, like the one our economic and political system currently depends on, values strong "bosses" and rules. But thinking about leaders as "bosses" disempowers most people trying to get involved. Leaders give us someone to blame (which then lets us affirm our own innocence), but they can't solve our problems without us. So we suggest that leaders are best understood as *conveners*—as people who bring others together. Unlike bossiness, leadership is not a personality trait or a matter of style. It therefore requires no specialized training (although, as with all skills, practice is helpful). Quiet and shy people can make great leaders. Leaders *create the conditions* for people to engage each other and be their full selves. Make the container for others to fill. A leader's job is to set up a good space and draw out good conversations and efforts. In other words, the job is to facilitate. You can have several leaders, or rotate, or even look to leaderless organizing models like direct democracy.[11]

Small groups for the win

Even within large associations, small groups are the unit of transformation. A group of two to five is little enough for shy people to feel comfortable, and for conversations to have real give-and-take. In a small, safe container, people can reveal extraordinary vulnerability. We bridge our individual experiences with those of the bigger group through these clusters. So—in whatever scenario—if you're feeling stuck, start with small group activities or conversations.

We are citizens, not consumers

A customer shops for prefab options, accepting (or rejecting) what's in front of them. The customer's only powers are purchase or complaint. When we view ourselves as customers or consumers, it impoverishes our relationships and our imaginations. Being a citizen (and here we mean citizen of the Beloved Community; this status is not dependent on one's documentation in any country) is a much bigger identity, based on taking shared responsibility for a future that we want. As citizens, we give our gifts freely and gratefully accept the gifts of others. This practice is not a transaction. We affirm our power, rather than surrendering it to others. We see the truth of our interconnectedness: we rise and fall together.

Process and outcome are linked

In other words, how we work with each other matters to what we achieve. If we want an inclusive society, we practice inclusion in our small communities, enduring the discomfort that arises when we do something new. If we want a world in which women participate in activism, for instance, then we welcome children into our activist spaces. This doesn't mean that we must be perfect, or that our small communities can fix all the ills of our large society by being internally flawless (and indeed, we'll never be flawless). But it does mean that we enact our ideals as we pursue them.

Ask big questions. Answers are less important

Questions are more powerful than answers, because they require us to consider the future. For instance, we start thinking about what sort of world we want when we ask ourselves what it will look like for our kids. Seeing this predicament clearly affirms our humanity—paradoxically, it's the lack of an easy answer that moves us to action. Powerful questions are often ambiguous, personal, and invoke some sort of anxiety; asking them lets you see what's in your heart. Solutions to powerful questions don't just fill potholes—although they can—they also deepen our understanding and commitment.

Welcome unlikely bedfellows

When we're explicit about our values and goals, we might find that we have quite a lot in common with people we thought were our opposites. Rev. Dr. Davies, the United Church of Christ minister we met earlier, told us about an improbable partnership she recently developed with another community in Ohio. She's been collaborating with a coven of neopagan witches in Cleveland to teach classes on how to create rituals. Rev. Dr. Davies began the project because she was exploring ritualism with friends; she started writing rituals for people and posting them on Facebook. "I noticed that the non-churched people were so much more attentive to them," she says. That is, people who didn't attend church still hungered for ritual, and had no blueprints for how to create it. So she now works to help pagans connect to ritual, in ways that are *broader* than her church, but that draw on a shared vocabulary of spiritual practice.

And while we're at it, rituals—broadly defined—shape a collective identity, and help us feel that we belong. Notice what rituals you already observe, whether it's a daily prayer, a weekly chess date, or regular pizza for phone-bankers. Invent new rituals as needed: a song for marching or a game to welcome new members. For example, Meghan's band asks new members to make an animal sound after their first rehearsal; existing members have to guess what the animal is. Ritual objects can also be helpful; several of the activists we talk with in the next two chapters are entrusted with their group's own megaphone to bring to demonstrations.[12] One megaphone in particular is clearly sacred, shrouded in stickers, anointed in spray paint, and bestowed with the thousand dings and scratches that mark a useful life.

Share responsibility

We are most accountable and committed to things that we help create. If we make decisions together, we are more likely to honor them and work hard to see them through. People who have a hand in designing or building a public park are more likely to feel that it meets their needs and use it accordingly; if we grow food our neighbors like to eat in our community garden, they are more apt to help water it regularly. Entitlement is asking what others can do to create the future *for* us. Accountability is *to be mutually responsible for the whole*: it values collective intelligence over hyper-specialized expertise.

If there is a single theme in the values we named, it is to move from individualism to interdependency. Making this shift is crucial as we join the climate struggle. We can see the dangerous failures of boss-style leaders all around us. In the following chapters we'll describe great work that's already being done across the country; from the municipal to the federal level, from inside and outside the halls of power. We'll break down the urgent work that needs doing: reforming, resisting, recreating, and reimagining. None of it is possible alone.

By organizing ourselves, we can move politics into the realm of public service and ditch our dependence on professionals and authorities in favor of a practical faith in our loved ones. Together we can move from fear to hospitality, wherein we meet strangers with welcome rather than wariness. We can value belonging over instrumentalization. Instead of focusing on our deficiencies, we allow ourselves to be whole.

LOVE IN ACTION

Jennifer Boylan, who we met earlier in the chapter, became active with Moms Demand Action for Gun Sense in America. Her story is an account of how community can bring us from feeling isolated and lonely to connected and powerful. Jennifer did not consider herself an activist for much of her life. But, as a young mother, she watched a rash of school shootings with deep pain and fear. "I was leveled by Sandy Hook," she says. "I had no experience with any of this stuff. I was just a mom, with little kids going to school. I was *leveled.* I basically cried for weeks. And when everyone seemed to have moved on, I was still hung up on it."

Jennifer's grief propelled her to her first action, which she took alone. "I did what I thought Americans did," she said. "I heard there was a march in DC, so I bought a ticket and went to DC. That's how I thought Americans changed stuff."

After the march though, nothing had changed. Sandy Hook was followed by the Charleston shooting, and by dozens more. But when she got home, she joined the new-ish Facebook group that would become Moms Demand Action. Through it, she met other organizers in Rhode Island, and learned to lobby. "I got involved in going to the State House," she says. Connecting with other people in her area really moved her. "To find an organization of countless Type A women like me?" she says, laughing. "Women who are going to get stuff done? We're not a bunch of wishy-washy people! And to have them all collected in one organization, coast to coast, in every state and every town. It was literally *my people.*"

For her, the lessons are long-lasting. "This organization has empowered me to get off the sidelines and actually effect change," she concludes. "I can't sit around and say 'someone should do something,' unless I'm willing to do it. It's gotta be me. And I'm so grateful that this organization took me—and lots of me's, people like me—and taught us what to do, and gave us the tools and resources we needed to literally change things. The lesson is, I think, that anyone who really wants to can roll up their sleeves and find a place to do this work."

Jennifer's organization provided her with the training, tools, skills, and community support to throw herself into the fight. However we define success, whatever our needs, we begin by forming and deepening relationships.

CHAPTER 7

The Big No

WHAT IS THE BIG NO?

We need to stop a ton of life-threatening stuff that is happening right now. The Big No is how we object, how we screech to a halt, how we make the oil stop flowing, make the smoke stop billowing, how we take money and social license away from the executives and politicians that are selling us out. The Big No drags hidden harms out into the public square and exposes them to the light. The Big No is anti-fascist. It says, "if your solution isn't seeking justice, it isn't a solution." It says, "there will be trials for this." It says, "we are done accommodating a fantasy that things can stay the same." We need Big No strategies for a life-and-death struggle.

As journalist Amy Westervelt put it, "the window of opportunity for moderate solutions on climate was slammed shut in favor of oil company profits for 30 years and now all of the options are extreme."[1] The Big No understands the differences between those extreme options with clear eyes: we know dangerous, violent ideologies are gaining momentum, and we say No to closing borders and dehumanizing refugees, immigrants, and internally displaced people, No to population control and anything less than reproductive self-sovereignty, No to billionaires, gaping income inequality, rent stress, foreclosure, student and medical debt, unemployment, houselessness and all the specific bad outcomes that dysregulated capitalism wishes to pass off as immutable facts.

But in the face of so much harm to stop so quickly, picking where to start can be overwhelming. We have four guiding principles to help. The Big No is:

1. Bigger than you.

2. Inspiring to you.

3. Powered by a vision of a life worth living.

4. Possible. That is: a project, not a general idea.

So what *isn't* the big no? The Big No is not NIMBY (short for "not in my backyard). When we say the Big No, we're saying "not here, not ANYWHERE." When we oppose something locally, we're opposing it because it's harmful, and will be harmful anywhere. If it is harmful to *our* bodies, it is harmful to *everyone's* bodies. What we say no to substantially defines who we find common cause with, perhaps even more so than our Big Yes. We can't build a future together if *I* excuse someone who is harming *you* today. Nor if you condone harm to me. The big No builds solidarity.

The Big No is disruptive: it doesn't pretend that good technology and minor modifications will save us. More wind and solar energy production means nothing if coal and natural gas plants aren't shuttered simultaneously. An eight-lane traffic jam of electric vehicles is not a win. Our ambition is to interrupt the status quo together, and build a better one. Therefore, we practice saying "No" on a heroic scale.

The Big No is not easy. It is often awkward, like shouting "the emperor isn't wearing any clothes!" It can be obstructing something that others enjoy, opposing something others want, naming something that people have tacitly agreed not to talk about. It's rarely a popular, polite, or easy thing to do. But it can certainly be liberating, both to the people saying no, and to everyone who was hoping someone would say what they're thinking.

Why are we bad at it?

Saying "No" can be especially hard for women. There is a special violence that's reserved for women who are perceived to be transgressing social boundaries of likability, and saying "no" clearly does just that.[2] A 2013 study published in the *European Journal of Social Psychology* found evidence of widespread prejudice in the United States against environmentalists and feminists.[3] This internalized prejudice is powerful enough to deter some people from action. Our generation has watched ecosystems collapse in real time while inequality and bias were shaping all our lives. But we also watched those who spoke up caricatured as hysterical naysayers—especially women, who are socialized to be accommodating and nonconfrontational. Many of us recognize the shorthand "crunchy granola," and "tree-hugger," and the more malicious, and gendered, "man-hater," or "feminazi." From the 1990s until the late 2000s, the default cultural posture toward activism was one of ironic detachment and cynicism.

Since the 2008 crash and the following wave of organizing that included Occupy Wall Street, activism has gotten both cooler and more necessary in public opinion. But today's adults still grew up with this cognitive dissonance: on the one hand an existential threat and widespread injustice; on the other, a powerful cultural disdain for making a scene, speaking sincerely, or making "unrealistic" demands. Many in our generation are still struggling against this psychic roadblock. But as Elizabeth Kolbert puts it, "any genuine preparedness strategy must include the averting of eventualities for which it is impossible to prepare."

This chapter draws on the insights of seasoned activists to offer strategies for finding our places in organizing and prioritizing our own work and thinking. We are learning how to be more obstinate and principled, to protest what we know is harmful.

How do we learn to do it better? Practice!

So what does saying "No" mean in practical terms? What must be stopped, and what do movements to stop those bad things look like? How do we make our No big enough? For a habitable future, we'll need to unseat corrupt and backward-looking leaders, oppose all fossil-fuel infrastructure and exploration, and end the economic policies that allow people and "people" (corporations: Thanks, Citizens United) to profit from the exploitation of natural resources and each other.

In this chapter we'll introduce some activists whose work has been primarily focused on stopping harm and/or rejecting the status quo. We also profile some successful and/or ongoing campaigns and projects that might light up brains in unexpected ways. All of them are collaborative actions. We chose them to give you a sense of what's possible, what's already afoot, and the breadth of tactics and tools available to us when we practice the Big No.

Background person power

Even though it takes bravery, the Big No is not about heroes or celebrities. The United States has a treasured cultural myth about the activist hero, in which social movements are simply vehicles that carry lone, great people (usually men) to triumph over evil. In real life, these great people understood the power and limits of their figurehead positions, and would often be the first to tell you they never accomplished anything alone. But still: Indian independence is frequently summarized by a quote from Mohandas Gandhi, the story of the decades-long struggle for African American civil rights is distilled down to a clip of Dr. King; Harvey Milk becomes the lonely figure of gay liberation.

This myth harms us in a couple of ways. One is that it creates an unattainable standard: *because that's not actually how social movements or protest actions work.* There is always a community.

There's always a planning team, even if the action has a single public face. A huge amount of group effort goes into every action, and if the community bonds are strong and the action is successful, it's back to work, after celebrating.

The other harm that the hero story does is to paint activism as vain, selfish, performative, and ego-driven—something people only do to become famous. We know that people do everything they do for lots of reasons, only some of which could be considered pure or noble. Ideally, Big No activists assess their action plans based on their probable outcomes. It's not wrong to want to be acknowledged or even praised for sacrifices, for bravery, for any good thing we attempt or accomplish. And it's also not wrong to be a "behind the scenes" kind of activist, who may not want their role to be in the public eye. In fact, a useful way to picture the structure and cycle of actions for activist groups is like the life cycle of a mushroom: under specific conditions, for a few days, a mushroom is visible—eye-catching even—but most of its life, and most of its structure, is underground, hardworking and far-reaching, in the form of mycelial networks that ferry nutrients and support across the organism.

Josephine first crossed paths with Juliste Gogolinski in May 2016 when Juliste led a nonviolent direct action (NVDA) training in a church on the South Side of Chicago the night before a national day of action called Break Free From Fossil Fuels. People were arriving from across the Midwest to rally and march in Whiting, Indiana, site of the largest oil refinery in the region, the next day. In Josephine's words:

> There were more than a hundred of us gathered in the church's canteen, deciding whether we would risk arrest, most of us for the first time, as part of the protest. It was late, and Juliste had obviously come from a long day of work. Years later I teased her that her vibe that night had been One Giant Eye Roll, but equally striking was her matter-of-fact compassion. She

communicated clearly throughout that she understood this was a big risk and that people were afraid and that was a normal and healthy response. She also pointed out for those of us that hadn't yet understood, not everybody's risk was the same and not everybody was in the same place to take it. The kid on a student visa has more to lose in this action than their retired professor. The gender nonconforming person and person of color will be at much higher risk in a police environment than their cis, white partner. Juliste didn't valorize the risk, or encourage anyone to participate. Rather, she held the space for people to role-play, ask questions, and think it through. The next day, thousands of people marched from the shore of Lake Michigan to a BP refinery, where forty-one of us were ultimately arrested while blocking the gates to the refinery for about an hour.

We met properly a few years later while preparing to protest JP Morgan Chase's financial support of Enbridge—the company behind the Line 3 Pipeline—with a sit-in at a Chase Bank Lobby. This was a smaller group, and I asked Juliste to teach me how to offer jail support. The night we spent in the lobby of the 1st Precinct police station, playing twenty questions and waiting for three brave arrestees to be released, was the beginning of our friendship.

Juliste is thirty-eight, trans femme, and grew up between the west side of Chicago and a central Illinois mining town, by way of Kraków, Poland. She calls herself a reluctant worker and a community organizer who stumbled into being a direct action trainer. She works as a union carpenter and considers herself a "background person" in the intersecting Midwestern movements for justice. She sat for an interview this fall in her backyard garden, which had been bedded down for winter. Her chickens were murmuring and scratching in the background of the recording of our conversation, joined by grinding car traffic and the occasional siren.

Juliste understood the life-or-death power of activism as a kid:

I knew I was trans and I knew I was queer at a very, very young age. That was the height of the AIDS epidemic and I saw the type of activism folks at Act Up were doing, or I heard about folks I'm related to going on strike, or . . . family members being laid off, factory closures. . . . [I] realized pretty young we had a fight just to, like, not die from a virus and be able to afford a roof and food and stuff like that. So, you know, I was just very eager. I was showing up wide eyed, this big sponge that wanted to just be around and take part and, like, do the thing. I was a little hang-around, being like, "Can I do this? Can I do that?"

And the direct action element really spoke to me, the troublesome aspect, the prankster aspect of it. The confrontational aspect really spoke to a lot of the internal anger I had at the time. And I saw it as a good channel for it. I saw it as a way that I could constantly grow and get creative and stuff.

But despite her appreciation for the power of confrontation, many of the roles she's taken on have been behind the scenes: building lockdown equipment, training participants, photographing actions, generally helping things go smoothly, safely, and well. She concludes:

As much as I climb this thing and block this thing and lock myself to this thing, I'm out knocking on doors too. I'm sitting through these community meetings as well. And a lot of times I'm literally showing up with saws and hammers and building things. There's so much that goes into everything, like pet setting and child care, and how are people going to eat? What are people's medical needs? On and on. My favorite strategy is the one that's gonna stop the bad thing from happening. Stop the harm as long as we can. It's going to build community. I don't think direct action leads to much without serious organizing. My favorite strategy is when all those things combine.

Nonviolent Direct Action

The Big No can be satire, provocation, barricades, or policy. It can be legal, illegal, or contain elements of both. Although much No can be accomplished through official channels, the classic Big No is direct action, which takes place outside of institutions, often in the streets. And as we mentioned earlier, research shows that nonviolent direct action campaigns are twice as likely to achieve their goals as violent campaigns. Today the most familiar forms of nonviolent direct action are marches, strikes, boycotts, and media/social media campaigns. But the language of NVDA is actually much richer than we often imagine.

The political scientist Gene Sharp compiled a list of 198 Methods of Nonviolent Action, which he divided into the following headings and categories:

- PROTEST AND PERSUASION—*Formal Statements, Communications with a Wider Audience, Group Representations, Symbolic Public Acts, Pressures on Individuals, Drama and Music, Processions, Honoring the Dead, Public Assemblies, Withdrawal and Renunciation*

- SOCIAL NONCOOPERATION—*Ostracism of Persons, Non-cooperation with Social Events, Customs and Institutions, Withdrawal from the Social System*

- ECONOMIC NONCOOPERATION: ECONOMIC BOYCOTTS—*Actions by Consumers, Action by Workers and Producers, Action by Middlemen, Action by Owners and Management, Action by Holders of Financial Resources, Action by Governments*

- ECONOMIC NONCOOPERATION: THE STRIKE—*Symbolic Strikes, Agricultural Strikes, Strikes by Special Groups, Ordinary Industrial Strikes, Restricted Strikes, Multi-Industry Strikes, Combination of Strikes and Economic Closures*

- POLITICAL NONCOOPERATION—*Rejection of Authority, Citizens' Non-Cooperation with Government, Citizens' Alternatives to Obedience, Action by Government Personnel, Domestic Governmental Action, International Government Action*

- And NONVIOLENT INTERVENTION—*Psychological Intervention, Physical Intervention, Social Intervention, Economic Intervention, Political Intervention*[4]

When brainstorming, it's a wonderful document to read in full. Sharp organized it in 1973, so there's a notable absence of digital strategies on the list, but it's obvious how immediate worldwide communication can accelerate and amplify many of the listed methods.

One specific note about boycotts here: Sharp lists twenty-five distinct types of boycott. There are many present-day examples of active, coordinated boycott campaigns, but there is a neoliberal tendency to think of a boycott as just *not buying something*. In the last chapter we talked about our transactional relationships and making the most out of "voting with your dollar." A boycott is a *coordinated* campaign of financial and public pressure, a political possibility that has been turbocharged by the internet. We haven't yet realized the full potential of a climate-inspired boycott, although we offer two examples of economic Big Nos that follow the same logic.

We often underestimate the "nonviolence" of NVDA in its strategic effectiveness. Phrases like "passive resistance" can give the impression that peaceful protest is about abstemiousness or saintlike conduct. Erica Chenoweth (whose research revealed the magic number for peaceful regime change: 3.5 percent of a population) points out that the "passive" in passive resistance is part of these strategies' success, because it removes a barrier to participation: only a small slice of people can or will join a militia, but "everyone is born with a natural physical ability to resist nonviolently. Anyone . . . who has kids knows how hard it is to pick up a child who doesn't want to move, or to feed a child who doesn't want to eat."[5]

There is also, obviously, a principle involved in nonviolence. That is, we don't harm *because we are trying to stop harm*. Nonviolence helps justice-seeking movements maintain the moral high ground, which is both essentially and strategically important to limit the number of possible criticisms your opponents can use.

But one of the most interesting, and risky, functions of nonviolence, especially in actions that include public confrontation, is that it can draw your opponent's violence out of hiding, *creating a public performance of the bullying force they try to conceal*. When documented, these violent outbursts can instantly turn a protest into a public morality play. There are many historical examples, but one of the most egregious in recent memory is from November 18, 2011, when during an Occupy campus sit-in, UC Davis police officer Lt. John Pike strolled down a line of seated student protesters, shooting pepper spray across their faces like he was watering a lawn. For witnesses around the country, Officer Pike's casual body language and equally casual violence expressed—much louder than words—state disregard for the well-being of its young.

But withstanding physical violence is an extreme version of NVDA, and only one method of many.

"I see direct action as literally *direct action*," Jayeesha Dutta tells us. Jayeesha is a lifelong activist and organizer, currently living in Port St. Lucie, Florida. When we asked her how many hats she wears, she laughed ruefully: she's the program director at the Windcall Institute, and a board member with I Witness Palestine, South Asian Americans Leading Together, Climate Justice Alliance and Alternate Routes. In 2016 she cofounded Another Gulf Is Possible, a climate justice collective, with members from across the Gulf South. "Taking direct action doesn't necessarily mean that you have to be blockading anything or being on the street protesting anything," she continues. "Taking direct action for me means doing something in direct solidarity to the issue that you're caring about."

So in the event of a climate disaster, direct action might look like mutual aid. "It might look," Jayeesha continues, "in the case of some food-related crisis, like growing food or harvesting and bringing it, right? Direct action, to me, is just taking action and doing something when you see something not right."

And indeed, from the beginning of the COVID-19 pandemic, Another Gulf Is Possible created some of the most useful, thorough, locally informed information and mutual aid resources in the country. Yudith Nieto, who we introduced in chapter 2, is another founding member of Another Gulf. She and Jayeesha are close friends and artist-collaborators.

Yudith explained why an artistic approach mattered in the creation of those resources:

> As artists, we emphasize that we're right there in the forefront with all the organizers. We're organizers ourselves. When you have an artist as an organizer too, that makes all the difference. You have your narrative strategy, and you have a much more beautiful and clear visual strategy of where the community wants to go, and what their solutions are. When we've gotten to create these things for community mobilization work, that has been the strongest way that we can contribute, and people appreciate us for it. And then they actually seek to work with artists and uplift young people becoming the artists of their communities, because now they know that they have a person on their team that can put something together: a banner, a visual, a brochure a zine, or infographics, which was a way that I contributed my artistic skills, because I am a trained graphic designer.

Her description of the virtuous cycle of arts and mentorship was also a good description of Yudith and Jayeesha's friendship, which they both told us about in glowing terms. Jayeesha tells us, "I remember first meeting Yudith . . . she was really young at that point. And I was just so impressed. I was like, 'This person is amazing.'" A few weeks later, when Yudith spoke with us, she noted that "Jayeesha is definitely one of my biggest teachers. She's an older sibling and we've learned a lot from each other. And I can say that now; she says she's learned a lot from me as well. So that feels good, because I'm not just the kid that hangs out with her homie. Now we're colleagues. And we really support each other."

We're dwelling, for a moment, on their friendship to offer a reminder from the previous chapter about what inspires people to take risks: trust, affection, and admiration for each other. We asked all our interviewees about their favorite memories from actions, and Yudith shared a funny—though necessarily vague—one. "We shut down this horrible conference of oil tycoons from all over the world," she says. "Somewhere sometime, we disrupted something and created a ruckus. For like 30 minutes we shut down their lobby, and were just loud and obnoxious to them and interrupted their good ole boy gathering. And that was really fun."

The public reaction was not altogether positive, however: "People had criticisms," she says. "But I think the proudest moment was that we felt in our hearts [something] needed to happen. And we did it. And we were proud of it. Even though it might not have had a big impact, they probably still talk about Darth Vader interrupting their cocktail hour. And it was just a message like, 'we're watching you. We're right here!'"

This story points to another important feature of direct action: we cannot know what our actions will achieve, or if they will succeed by the standard of our intention. This is the heart of our active hope; that all outcomes are unknown. But we also know that if we're resorting to NVDA strategies, it's because we're outmatched in *every other way but people power and vision*. The fossil-fuel execs have money, connections, infrastructure, cultural dependency, conventions, and habits, and frequently the laws on their side. Each specific action we take can feel like, and frankly, *be* a failure.

Cherri Foytlin, another Gulf Coast activist and artist, published a zine in 2022 called *The Enchiridion—Failing to Save the Earth, a How-to Guide.* In it she advocates creating a Failure Plan, which she insists is not planning to fail, but *planning to succeed regardless of outcome.* Her steps for creating this plan are:

1. Identify the values that underlie your concern. Why is it important? (i.e., to protect the water)

2. Map the entire depth of the threat(s) to those values.

3. Contact intersecting stakeholders, efforts, and mutual opponents. Build trusting and genuine relationships.

4. Collectively design "failing" objectives (i.e., even if we do not stop the pipeline, we will raise awareness to 5,000 of our neighbors).

5. Think beyond this battle. What efforts would best serve your values, and move the movement forward into the future?[6]

It's worth pointing out that any time your opponents spend dealing with you is time they *can't* spend on business as usual. The more interruptions, and different kinds of interruptions, a movement can generate, the more of their opponent's resources they can monopolize. Any specific action can feel like, or *be* a failure, and still succeed at "failing" objectives.

VARIETIES OF BIG NO

There are as many different ways to say No as there are people to say it. And as the work of Another Gulf Is Possible shows us, distinctions between what's against harm as opposed to for resilience, or between what's cultural as opposed to political action, aren't always clear-cut, or important. But here we'll give a bit of structure to the beautiful chaos of climate resistance by breaking some recent examples of Big No into four major categories: Social/Cultural (bad culture), Political (bad laws), Infrastructural (bad infrastructure), and Economic (bad money). Activist groups are tackling each of them with creative "No" tools such as educational campaigns, lobbying, public pressure like naming and shaming, theater and protest music, boycotts and direct action.

For chart enthusiasts, see figure 7.1 for one way to understand different points of intervention.

	Municipal or Local	State or Regional	National
Socio-cultural	• Sit-ins, walkouts and demonstrations • Op-eds/letters to editors/cartoons • Strikes and slowdowns • Street theater • NVDA trainings	• Info/resource-sharing networks • Coordinating local campaigns • Writing letters/op-eds • Teach-ins	• Media/public education campaigns • Memes/hashtags • Holiday recognition (e.g. May Day strikes) • Deplatforming • Name and shame
Political	• Residential gas ban • Traffic/parking tax • City sovereignty laws • Bird-dogging elected officials • Water use regulation • Pollution regulations	• Fracking ban • Ending fossil-fuel subsidies • Pollution lawsuits • Pollution regulation • Pesticide bans • Lobbying	• Ending fossil-fuel subsidies • Pollution lawsuits • Ending industrial agriculture subsidies • Ending *Citizens United*
Infrastructural	• Ban on fossil-fuel hookups in new buildings • Fossil-fuel infrastructure blockade • Oppose industry in residential areas	• Ban on fossil-fuel hookups in new buildings • Campaign against fossil-fuel infrastructure • Pipeline blockade • Forest and water protection	• Freight regulations • Public pressure against fossil-fuel infrastructure • Stop deforestation • Coal moratorium • Ban oil transport by rail • Forest and water protection
Economic	• Local divestment (school, city, etc.) • Pressure banks to defund fossil fuels (local branches) • Business boycotts • Oppose corporate-centered development	• Divestment of state pensions • Pressure banks to defund fossil fuels • Product boycotts • Millionaire's Tax • Equity impact analyses of all new laws	• Divestment • Pressure on bank headquarters to defund fossil-fuel projects • Boycott campaigns • General strike

Figure 7.1

Big No from the Heart

But before we get to specific targets, we begin with an imaginative exercise. Whether or not we know how to express it, many of us feel in our hearts that we are living wrong, that violence and destruction *must stop*. We may not be able to express this in the form of a specific, satisfiable demand, but the direct emotion can be even more powerful for its implacability. The heart knows that we must say No to be right with ourselves. Think of when you most feel outrage, exasperation, mutiny. The sheer massive badness of something can weigh down our No with fatalism, because we have no idea how to stop or fix it.

But sometimes just expressing the feeling is enough to get started.

One Big No from the Heart reverberated in the streets of Columbia, South Carolina, in 2015, when musician Matt Buck brought his sousaphone to a white supremacist march at the State House. He followed the Nazis, playing music from cartoons and a down-tempo "Ride of the Valkyries." His goal? To make loud, stupid sounds at loud, stupid racists, because "far too often, people get away with being idiots."[7] This is one way deplatforming can operate in the real world, in which a Big Farty No drowns out and delegitimizes hate speech. We may not all play sousaphones, but we can each make a big noise.

From a one-man band we look to the Occupy Movement, one of the largest recent examples of a Big No without a specific demand. Adults on six continents gathered in public parks (or, notably, privately owned "public" spaces) to interrupt the status quo and say No, this isn't working for any of us.

One version of history states that Occupy failed by ending, or by not having satisfiable demands. But being part of Occupy flipped a switch for many that hasn't turned off since. "It's hard to overstate the inhibition that has been squatting like an imp on the anxious chest of 'responsible' opinion for two decades or more. Call it the inhibition of realism," wrote Jedediah Britton-Purdy, reflecting on his liberatory experience of Occupy in Chapel Hill, North Carolina.[8]

Juliste observed something similar. "I think what was amazing about Occupy," she says, "are two big things: seeing capitalism as a harmful institution—an unsustainable death cult—was brought to the national conversation." She pauses while a neighbor starts a very loud engine. "And also [now, when I show up to] any kind of march, any kind of protest, I look around and I see people that were brand new to activism during Occupy, and I see them as like, legal observers and street medics and they're cooking the food and they're carrying the things. Like how many people are still involved in it and how many of those folks are union organizers now? How many of those folks are housing organizers? And when those big things pop up after the wave has died down, how many people stay involved in the work and disperse out into more pointed, more specific parts of movement work?"

Occupy was rooted in preexisting movements and branched out to strengthen many more. But regardless of where Occupiers ended up—whether or not they are union organizers or activists at all now—hundreds of thousands of people worldwide showed up to voice their No together, and were heartened by that experience. This cultural outpouring taught a generation of activists that when we start with a Big Public No, another world becomes that much more possible.

Social/Cultural
Skolstrejk för klimatet

The youth-led global School Strike for Climate, also called Fridays for Future (FFF), Youth for Climate, Climate Strike or Youth Strike for Climate, is a powerful example of direct action—from the heart—for cultural change: children refusing to go to school like normal to protest their Elders' lack of climate action. These are local actions with specific demands that vary by community, but they have added up to a mass, international movement. And while we see a lot of press photos showing crowds and rallies, if you follow the hashtags you'll also see a lot of lonely ones and huddled twos on various capitol steps. The demands of the larger movement are formulated around a justice-informed 1.5 degree goal, but the protests demonstrate a more powerful No than could be expressed in policy alone.

Like nonviolence draws violence into the open, these children's action draws inaction into the open for scrutiny. The narrative is that if adults had been saying No already, our children wouldn't be forced to. In the book *Facing the Climate Emergency*, Margaret Klein Salomon suggests, "If you're the parent of school-aged children, consider an organization that works to organize school climate strikes or works with students in other ways."[9] It's understandable that adults may want to encourage their children to participate in #skolstrejk as a way of expressing their own urge to do something. It has momentum, demonstrated power, and moral clarity.

But its power comes from its emerging directly from kids, not being puppeteered by grown-ups. Of course parents must support and encourage their kids' interest in striking, but *our own* task is to imagine the adult equivalent instead. Remember Dr. Susie Burke: "The best thing parents can do is actually just to get really busy themselves with their own climate action—and to be doing *collective* action—because children need to know that their parents care for them, and actually also care about protecting their world."

It's destructive to sentimentalize—and instrumentalize—children. Let's not fall into kitsch. "Kitsch causes two tears to flow in quick succession. The first tear says: How nice to see children running on the grass! The second tear says: How nice to be moved, together with all mankind, by children running on the grass," wrote Milan Kundera in *The Unbearable Lightness of Being*.

How sad to see that child so angry at parliament! How beautifully sad we all feel that the children will have to fix this broken world we've left them!

Child activism is often seen as "lower stakes" than a workplace strike, or occupying a congressional office. Yes, the law exposes adults to more serious consequences for direct action, and adults will also face fewer consequences of inaction. Focusing on kids' activism because it already exists, and is therefore easier, hides the ugly fact that it's occurring because adults have ignored our own responsibilities. We can find moral clarity at every age.

Climate Anxiety Counseling booth

Author Kate Schapira made a cultural intervention on the hyper-local level with the Climate Anxiety Counseling booth. She began in 2013 in Providence's Kennedy Plaza, with a Charlie Brown style table and sign reading CLIMATE ANXIETY COUN-SELING. 5 CENTS. THE DOCTOR IS IN.[10] Her project had spectacularly low overhead. It was interactive and compassion-ately inviting. Other amateur therapists borrowed her concept, made their own signs, and began having their own streetside conversations with strangers about their emotional experiences of the climate crisis. The effects of this intervention may be unquan-tifiable, but you can tell from scrolling through Schapira's "Booth Sessions" on her blog that her project initiated a huge number of raw, personal conversations with diverse members of the public.

Political (Bad Laws)

Much of what's heating the atmosphere is perfectly legal. Obvi-ously, that doesn't make it permissible. Changing laws takes different methods than changing public opinion, although these projects can bolster and further each other. Laws that protect the fossil-fuel industry and its profits have to be repealed, laws and regulations that protect people from pollution have to be passed. There are many kinds of campaigns to change laws.

Delaware River Basin Fracking Ban

Fracking, short for hydraulic fracturing, is a way to get gas— mostly methane, a greenhouse gas—and oil from underground shale beds. Shale rock is made of many thin, fragile layers of compressed mud and clay. Fracking companies drill horizontal wells, and then they force sand, water, and a mix of chemicals into them. That makes the shale crack and release oil or gas. As you remember from chapter 1, fracking poisons people who live near fracking wells, in part by contaminating drinking water.[11] Infants are especially vulnerable.[12] When methane is combusted, it releases carbon dioxide, another greenhouse gas, and micropar- ticle pollution. While fracking has occurred since the late 1940s, the United States has experienced a boom since roughly the turn of the century.

New York banned massive hydraulic fracturing by executive order in 2010, so all fracked gas production in the state is from wells drilled prior to the ban. Vermont banned fracking preventa- tively in May 2012. In March 2017, Maryland became the second state in the United States with proven gas reserves to pass a law banning fracking. On May 8, 2019, Washington joined forces with the other three states. Here though, we focus on a particular region: the Delaware River Basin. The Delaware River provides drinking water to more than 17 million people, and Delaware has shale reserves, making it a desirable spot for gas companies to set up fracking operations.

In 2010, the Delaware Riverkeeper Network and partner organizations secured a total moratorium on all aspects of the fracking industry within boundaries of the Delaware River Watershed. That included the actual fracking operations, as well as the import for treatment, storage, or disposal of toxic frack wastewater, as well as water exports that could allow water from the basin to be used in fracking elsewhere.

In an historic move in 2021, the Delaware River Basin Commission—the body responsible for safeguarding its water quality—formalized this moratorium in the form of a ban, voting to outlaw fracking in the region. Particularly noteworthy was that the governors of New York, New Jersey, Pennsylvania, and Delaware all voted for the ban. That vote, and the ban, was the culmination of work by the Delaware Riverkeepers network and dozens of partners who waged a comprehensive, years-long campaign. From publishing opinion pieces and blogs, to circulating petitions, to direct actions, to mass attendance at Commission meetings, this coalition pushed relentlessly against the poisoning of their lands by the fossil-fuel industry.

While this ban was a major victory, organizers were hoping for a total, and permanent, ban without any loopholes.[13] Now, their focus is shifting to banning the import and export of fracking wastewater. Riverkeepers worry that the fracking industry will take advantage of loopholes in the regulations. Such loopholes could result in toxic wastewater being stored on local land, which would allow evaporation of those toxins into the air. Or the fracking industry could use the wastewater in other harmful practices like concrete production.[14]

Community Environmental Legal Defense Fund

One group that is working to prevent and close this kind of loophole in community legal rights is the Community Environmental Legal Defense Fund (CELDF). They offer legal support to towns who want to push back against extraction and pollution, and enshrine Community Rights and the Rights of Nature in local law. Their big picture plan is to build the local legal foundations to challenge damaging state and federal laws. In their words, "we recognized that whether communities were facing fracking, injection wells, factory farms, pipelines, greedy bosses, landlords, GMOs, water extraction, or a wide range of other threats, the barriers they faced to stopping these projects—and in their place establishing sustainable energy, water, agriculture, and other systems—were the same."

Their years-long collaborations with towns in Ohio, Pennsylvania, Oregon, New Hampshire, and Maine have led to fracking bans and water protection laws, as well as laws that protect the right to public protest. They have also successfully protected townships against lawsuits filed by polluters, and opposed "corporate rights" through local ordinances.

Infrastructural (Bad infrastructure)
Power Past Coal and the Thin Green Line, Pacific Northwest, 2012–2016

The movement against coal export in the Pacific Northwest was Josephine's introduction to the world of climate campaigning. She writes:

> When I moved to Olympia, Washington, in 2012 so my partner, Chris, could attend the Evergreen State College, I was looking for clues about how to start actually doing something useful for the climate. I had no idea, but saw that Bill McKibben was putting on an event in Seattle, a stop on his "Do The Math" tour. I dragged Chris and two friends with me, and signed up for all the mailing lists with activists in the lobby of Benaroya Hall as the event let out. In the next week I got an email from an organizer named Robin Everett that began:
>
>> Thank you so much for the concern you have shown in the fight to keep Big Coal out of the Northwest! As you heard at Benaroya, the Seattle public hearing for coal exports has been moved to Thursday, December 13th from 4–7 PM (Rally at 3 pm) at the Washington State Convention Center, in downtown Seattle. That doesn't leave us with much time, and there's still a lot to do.
>
> I told her I wanted to help but lived in Olympia, so she sent my email to Beth Doglio. Beth called me about 5 minutes later. She

was running the Power Past Coal campaign out of her office at Climate Solutions—a regional policy advocacy nonprofit—and she did not have a lot of volunteers. Things moved quickly from there: my to-do list from that week included Send Phone Bank Invites, letter to The Olympian, phone bank Dec 11 (also Dec 12?), Rally Dec 13. Beth asked me to testify at the hearing after that rally, and I did. We were speaking to the agencies whose job it was to decide what kinds of impacts they would consider when they studied the proposal for a new Coal Export Terminal at Xwe'chi'eXen, also called Cherry Point, Washington. I didn't know anything about this, except the broad strokes that we need to stop digging, selling, burning coal, and that dredging inlets and building infrastructure on Lummi land against the wishes of the Lummi Nation was clearly wrong and illegal. It was okay that I wasn't an expert; the point of the public hearing was to have community input in the process. Beth helped me structure my comment correctly for the process, but the inexpert opinions were all mine:

"NEPA [National Environmental Policy Act] and SEPA [State Environmental Policy Act] should investigate the following issues for the DEIS [Draft Environmental Impact Statement]:

- The Direct Adverse impact on our atmosphere burning this coal will produce. There is no local or foreign atmosphere. If we ship and sell the coal we speed the damage to our climate here and abroad. My family lives in New York City and has just survived an unprecedented weather event [Hurricane Sandy] caused by too much atmospheric carbon. This significant impact cannot be mitigated if the proposal is approved because its purpose is to release carbon.
- In the face of this climate crisis, if we're considering the employment of rail workers to be a significant benefit of this proposal, a more sensible alternative would be to expand and update national passenger rail infrastructure, to get people out of cars and airplanes.

- Also it's clear to me at twenty-nine years old that we
  have a precious few years before it's obvious to everyone
  that burning fossil fuels is an economic liability, not
  an asset, and the market will not want, nor be able to
  demand those commodities."

Not sure I would have written it quite like this today, but I stand
by the points. The reason I pointed out that I was twenty-nine
is because at that time in Olympia the median age for climate
activism was about seventy-five, and Beth was excited to have
a young face to take around. Even though public speaking is
not my favorite, I spoke at rallies and testified at a lot of public
hearings in those two years, in favor of climate legislation and
against other fossil-fuel-infrastructure proposals.

Beth helped me try on an array of activist hats, also putting
me in touch with an editor who wanted someone to cover the
Power Past Coal campaign, which became my next job. Beth
made the workings of the State House comprehensible to me.
It helped that during that time the State House was walking
distance from my house. It also helped that I arrived toward
the beginning of a campaign which began in opposition to
one proposed terminal and grew to shut down upward of nine
similar proposals from British Columbia to Oregon, with
connections to anti-mining campaigns as far away as Alaska
and the Tongue River and Powder River basin coal seams in
Montana and Wyoming. The international coastal alliance of
climate activists has come to be known as the Thin Green Line,
across which fossil fuels shall not pass. Their work is ongoing, as
they oppose fracked gas and tar sands oil pipelines, the increase
of oil and coal by rail (with the attendant increase of hazardous
derailments), and zombie coal projects.

On our way home from tabling at a green energy event, I
remember Beth telling me how her organization, Climate
Solutions, came to run the Power Past Coal campaign in

Washington. She said that it was very unusual for them to take a "no" position against anything, that they focus on positive, often incremental, legislative goals and things that we *can* do, instead of things that we mustn't. But I'd been shoulder to shoulder with her, watching her wield considerable might against coal: a radical "No" without a conciliatory "Yes" in sight. For Beth it came down to integrity: "how can we trifle with appliance efficiency standards and call ourselves a green state when we're going to ship hundreds of millions of tons of coal to Asia?"

As we write this, Beth has just won her third term representing the 22nd District in the Washington State House, on a platform that includes the Green New Deal, universal health care, and protection for reproductive rights.

Economic (Bad Money)
Economic campaigns focus on money as a proxy for fuels, understanding the flow of finance to determine the movement of oil, coal, and gas.

#Divest
Probably the most famous form of economic climate activism is the fossil-fuel divestment movement. The divestment movement is a network of campaigns aimed at universities and other large organizations, pressuring them to stop investing in fossil fuels as part of their financial and/or retirement portfolios.[15] Divestment is a bottom-up movement, initially mostly student groups, although today there are groups working to divest everything from town pensions to churches and temples, and notably, the Funeral Consumers Alliance of Maine.

Protest divestment is a form of dissent in which stockholders intentionally sell their assets from a corporation to express dissatisfaction with the company's behavior. It took its inspiration from the anti-apartheid divestment movement of the 1970s and 1980s

(Apartheid was a legal and political system of racial segregation in South Africa, in which the National Party used racialized violence to uphold political and economic control of the Black majority by the white minority). Beginning in the 1970s, anti-apartheid campaigns in the United States and elsewhere gained momentum as violent repression of South African liberation movements intensified.

An international activist movement pushed businesses and institutions to divest holdings from South African investments in objection to the actions of the South African government, thus pressuring it to change its behavior.[16]

The same logic applies to fossil-fuel divestment. In 2013, Bill McKibben argued that college students should convince university administrations to sell off their investments in fossil-fuel stocks.[17] The strategy of divestment campaigns is to use stocks and investment portfolios as leverage to remove financial support from some of the most profitable corporations in human history—oil, gas, and coal. Divestment campaigners work within existing institutions to chisel away at the market share and power of fossil-fuel companies, intending to hit those companies where they'll feel it most keenly: in their financial solvency and relationships with shareholders. The purpose of this tactic is twofold: first, to take money and social license away from the fossil-fuel industry, and second, to stop schools and other institutions from profiting from climate pollution.

Rather than the individualized approach of "voting with your dollar," divestment activists use collective strategies like rallies, protests, and sit-ins to demand that administrations divest endowments from oil, gas, and coal holdings. While the goals remain market-centered, methods are collective, and designed to starve the oil and gas industry of its social standing, as well as its financial power.

Fossil-fuel divestment has been gathering steam in recent years; as of 2020, more than one hundred colleges and universities have committed to partially or fully divesting their endowments of fossil-fuel stocks. While the direct monetary impacts of divestment are still relatively small, the indirect impacts (in terms of public discourse shift, for instance) are significant. Divestment has put finance and climate change on the agenda of shareholder meetings and conventions of Boards of Trustees; it threatens the legitimacy, reputation, and viability of the fossil-fuel industry.

This threat has altered the finance industry; shareholders and investors are making new demands, and the notion of "fiduciary duty" is in flux. A fiduciary duty has typically been understood as the responsibility of a corporation to shareholders. But the divestment movement has challenged this. Fiduciary duty requires both the obligation of prudence and a duty of impartiality, meaning that companies must meet the needs of careful investors, but also not favor one party over another. Fiduciaries must put the interests of all their beneficiaries above their own interests and protect the assets of their beneficiaries equally. Financial analyst Kathy Hipple has argued that the volatility of the fossil-fuel industry, and its contributions to climate change, mean that (especially for young beneficiaries), "investments in oil and gas companies are almost impossible to defend," and that divestment is a moral imperative.[18] This was not a widespread spontaneous insight but the product of many people's ongoing activist work.

Many of today's influential organizers cut their teeth in college divestment campaigns. For McKibben and others, the divestment campaign is also an opportunity to build the collective power of a national movement; it is an opportunity to form new nodes in a network of mobilized campus groups. As a whole, this coordinated movement can push on many aspects of the climate crisis at once, raising awareness among college-aged youth, even as they force oil, gas, and coal companies to begin paying for the very real costs of climate change.[19] In this way, divestment groups may work within the system to expose the system, forcing the contradictions of capitalism into the light.

Stop the Money Pipeline

Growing alongside institutional divestment is a coalition movement to pressure banks to stop lending money to fossil-fuel projects. In 2016, as activists organized opposition to the Dakota Access Pipeline from the beginning of its proposed route, the Bakken Oil Fields of North Dakota, to its intended destination in southern Illinois—most famously encamping for months at the Sacred Stone Camp on the Standing Rock Reservation in South Dakota—some set their sights on the banks financing the pipeline expansion. Activists in South Dakota and allies around the world ran a sophisticated media campaign. Under the tag #Defund-DAPL, they focused efforts on SunTrust, Citibank, Wells Fargo, and others, who together put up close to $4 billion to lay pipe. Activists protested in branches and corporate offices around the world, and online activism mounted public pressure on banks.

Stop the Money Pipeline Coalition is a superb example of the "think beyond this battle" part of a failure plan. Even when Donald Trump expedited the permitting and construction of the Dakota Access Pipeline over widespread outrage in 2017, what efforts would best point the movement forward? The concept of a bank pressure campaign had already been tested. In the following years people put this strategy to work at countless intervention points with the shared aim of defunding fossil fuels.

Then in November 2019, by the Coalition's account:

> forty of the leading strategists from fossil fuel finance campaigns in the United States came together for three days in the woods of Vermont. The gathering was attended by leaders of the youth movement, Indigenous frontline leaders, campaign leads from efforts to hold banks, insurers and asset managers accountable, and representatives from some of the largest movement-building climate groups in the country.

STMP now includes more than two hundred partners. They offer grants, training, and run educational, direct action, and public pressure campaigns, all guided by the Big No idea that if money doesn't flow, neither will fossil fuels.

WHAT SKILLS DO WE NEED AND HOW DO WE GET THEM?

We pulled examples of interventions from across the country, at many different levels, and with many different targets. You may have a sense for the variety of skills that are needed by a movement this big and ambitious, and you may also feel daunted, as in: "what could I even do?"

Cincinnati-based organizer Jen Mendoza is a corporate campaigner by day. "I don't work with governments at all," she says. "I work on targeting corporations and building out a strategy to pressure corporate targets who are doing some of the worst. And outside of that, I've been a part of more off-the-cuff direct action for the last ten years." She spoke to us reverently about the Elder activists who mentored her. She is passionate about encouraging, and training, the next generation of leaders.

But first, and because she knew little about organizing, she read and reflected on her own connections. She recalled a heavy emphasis on building her constituency with the impacted people. "That's something that the environmental movement often fails at, because we are so issue-focused that we forget that there are almost always people directly affected by our issue," she says. "And I could point to literally every single movement and find the people who have a stake in what you're fighting for." She learned early to put people first.

Both finding experienced activists whom she admired and trusted and reading helped Jen find herself as an organizer and allowed her to make use of her natural talents and inclinations. She says her superpower is motivating and inspiring others: "we had truckloads of people from Cincinnati go up north for pipeline fights through me just literally getting on the phone and being like, 'do you want to be a part of this?'"

So what about those of us who don't believe we can move truckloads of people through sheer force of personality? "The people who I've seen plug in and stay were the ones who someone trusted enough to say here, go do this. And they went and did it and they felt great about it and they felt accomplished." This is exactly what Beth Doglio did for Josephine in the midst of the Power Past Coal Campaign.

There are also reasons why people leave activism: sometimes personal tragedies or health issues intervene, sometimes they find other channels for their change-making energy. But there is one big, preventable reason as well: not having a job. "They leave because they didn't have a role to fill," Jen tells us. "Someone didn't take the care to develop them as a leader and give them a significant role." This might feel like a small point, but it's a crucially important one: when we extend welcome, we have to make sure that everyone who wants a job has one, and that we make the best use of everyone's skills. Nothing is more deadening than being ready to pitch in but not being recognized.

We asked Jen what underappreciated skills she sees in effective activists, and she took a moment to reflect. "I think that in the secular Left, there is not enough space for spirituality, and self-care." She acknowledges that self-care can be a tricky subject because of the focus on individualism, and offers the antidote of community care:

> We have to have spaces where people can talk and unpack the human experiences we have with each other, with the land, with our theories and our movements. You did an action that required going to jail? Where is the space for people to talk about *what they saw in jail?* Where's the space for them to examine the *privilege* of going to jail when you signed up to do so, versus the people who didn't? If we are serious about creating

a world that operates outside of the destructive paradigms, then building up community is a crucial step we cannot miss.

For her, activism is about understanding what motivates other human beings, and helping to elicit their strength and what they care about. In general she believes the skill of holding space for the emotional parts of this work is wildly underappreciated. She expresses her pride in

> the people who do not miss an opportunity to pull someone aside and have a cup of coffee with them. Those who are emotionally intelligent enough to make sure that there are healthy, safe spaces for people. Those are the ones who we should think of as invaluable. That is the unseen work, the unpaid work, and it is typically the work of non-men. Those spaces are healing ones, and we have to stop pretending like we don't have time to create them.

Any long-term fight needs emotional and spiritual sustenance. We provide it for each other.

SIDE EFFECTS OF THE BIG NO

As Jen points out, climate organizations can act like we don't have time for the emotional dimensions of this work. This brings us to some of the pitfalls of "No"-focused activism. There is *so much* work to do: many lifetimes full. Once we wade in, it can be easy to keep going till we struggle to keep our heads above water. A total focus on stopping harm can lead to:

- A sense of playing Whac-A-Mole: saying No constantly can feel like ceding decision power to the bad guys and operating within their framework.
- Exacerbated misanthropy and human-hating.

- A distorted perspective of scale and one's own self-worth, either woeful smallness or exaggerated self-importance: "if I take a break the world will explode."
- Setting oneself up for disappointment by attachment to specific outcomes.
- Punishing oneself with unrelenting work.
- Forgetting to live, or to engage with what is loving and joyful.

When we offer the Theory of Balanced Climate Action, it's primarily as a reminder to keep an eye on the bigger picture: the values that guide us, the vision of a just, habitable future, and the knowledge that we may not know each other but we are in this together. Knowing that there is movement everywhere can make it easier to rest when you need it, and to find inspiration in others when you've misplaced your own.

And if or when saying No all the time becomes too costly, it's good to remember that a ton of Big Yeses are just as badly needed. As we've said before, the harms we work to stop and the world we want to create are joined at the root. In parting, Jen shared a story about the kind of moment that makes it all worthwhile. She had been mentoring a new activist, and they had set out for an action:

> And it was a tripod action [in which activists erect and climb on top of tall, wide structures to occupy public space or block traffic]. We were deploying two tripods. So we were in the back of a U-Haul with all of the poles and equipment. We were heading to the location, and there was a moment of nervous silence among like ten of us. Everyone was on edge. We don't know what we're going to come into. Are the police already going to be there? Everyone is really nervous and tense. And this person just started singing a song that everyone knows, and it's kind of a punk song, but a movement song. And I literally want to cry thinking about it, because it was just such a

powerful moment. To see them be so in tune with everyone and also with themselves and lead this song. There's a call-and-response part of the song. So like, *everyone* is now singing it. And it was just so beautiful. And at the end of the song, we were at location. Doors come up, and we just run out and do the action! And it was absolutely perfect. From beginning to end. It went without a hitch. And it was amazing.

YOU DON'T HAVE TO PRACTICE DIRECT ACTION, BUT YOU DO HAVE TO UNDERSTAND IT

We feel this chapter wouldn't be honest without one final piece. We are not all moved to direct action, to block traffic, to enact the Big No confrontationally. But we each have a minimum requirement: to understand and be sympathetic to these tactics, and to the people who are called to use them.

If, as Chenoweth's research tells us, we need to sustain 3.5 percent of the population as they engage in active disruption for system change, that small figure is rooted in a sympathetic majority at home, coordinating, maintaining, and supporting. Those are people who are not out in the streets, not necessarily protesting, but whose hearts, hands *and voices* are with the 3.5 percent, and not with the status quo. This support can be domestic and nonconfrontational, but it has to be explicit, out loud. Otherwise we're back to "pluralistic ignorance," the "phenomenon that occurs when people mistakenly believe that everyone else holds a different opinion from their own."[20] In silence comes implicit agreement: we're okay with the damage and harm.

A pernicious form of obstruction comes from people who are sympathetic to a cause *in theory*, but, when confronted with a real direct action and all the inconveniences, conflict, and blowback that it entails, disavow the action, if not the whole movement. Anytime there's a disruption to the status quo (a traffic blockade, a sit-in) there's a lot of ruffled feathers. And the feathers don't always just belong to the people who are the intended targets of the action. Public disruptions are often widely criticized, even as history proves them to be wildly effective.

We may find ourselves more upset than we expect to be when presented with other people's "misbehavior," feeling offended, guilty, or threatened. This can lead to the ol' kill-the-messenger routine: we know that climate change has a lot of different faces. It hides *itself*, as well as having been hidden by fossil-fuel industry–funded misinformation. So often, when activists drag harm out into public view, they are seen as *being* a problem instead of *addressing* a problem.

None of these critiques are new, either. Dr. Martin Luther King Jr. was scathing in his criticism of white, supposedly liberal, hand-wringing in the face of civil disobedience against Jim Crow. His position is worth quoting at length here:

> I must confess that over the past few years I have been gravely disappointed with the white moderate. I have almost reached the regrettable conclusion that the Negro's great stumbling block in his stride toward freedom is not the White Citizen's Counciler or the Ku Klux Klanner, but the white moderate, who is more devoted to "order" than to justice; who prefers a negative peace, which is the absence of tension, to a positive peace which is the presence of justice; who constantly says: "I agree with you in the goal you seek, but I cannot agree with your methods of direct action"; who paternalistically believes he can set the timetable for another man's freedom; who lives by a mythical concept of time and who constantly advises the Negro to wait for a "more convenient season." Shallow understanding from people of good will is more frustrating than absolute misunderstanding from people of ill will. Lukewarm acceptance is much more bewildering than outright rejection.[21]

Direct action is the language of life and death, not something frivolous or self-valorizing. Discomfort is part of it. Even if it's not a strategy that we are using ourselves at any given time, it's necessary to understand it, and be in solidarity with it at a basic level. We can critique the tactics of any action, *but we have to do it from a place of recognition.* That is, to understand that these are the tactics of people who find themselves outside the official channels of power, who are reaching for the biggest levers that they can, out of necessity. We all know that the climate crisis means life and death. Whether it's acute today for us or not, we know it's life or death for our children, our hopes for the future, our water, our earth.

So we acknowledge that it's life or death, even if some of us are experiencing a reprieve in consequences. That acknowledgment means we recognize the validity of direct action. If something feels distasteful to us, it's our responsibility to take a step back and hold that feeling up against our understanding of the big picture. If watching a public protest makes us feel so uncomfortable that we want to criticize the action or disavow the movement, it's time for reflection. If we call it counterproductive because some people will be turned off by it—*not me,* but *some people*—we need to reflect on that.

We can absolutely critique tactics: different direct actions are more or less successful, as are different methods. Some targets are more thoughtfully selected than others, some campaigns are better planned, or better staffed. An action may have better or worse optics and choreography. We can become part of decision-making groups that help shape the actions *we* think are most effective, and we can keep building the movement so that we can more easily recover and redirect from one unpopular action. But if we can't recognize and vocally defend any direct action's importance in the movement for a just future—however imperfect—we're siding with the status quo.

In 2022, a protest inside the National Gallery in London involved two activists throwing canned soup on the protective covering of Vincent van Gogh's *Sunflowers,* then gluing their palms to the wall of the gallery. This action was widely mocked and criticized, although one of its goals was to put climate change in the headlines, and in this it was very successful. We were surprised at how many climate-concerned people disavowed the action on the grounds that it was, basically, a turnoff.

One of the clearest articulations of understanding came from an unexpected ally, American audio engineer and musician Steve Albini. In a thread on October 14, 2022, the same day as the action, he schooled Twitter on how to understand direct action, even if it's not your thing:

- People are dunking on these absurdist demonstrations but I think they're kinda cool. The point they're making is that your most priceless shit is worthless on a dead planet, and while they're not going to change any minds it's a way to get that idea into the discourse.

- There's going to be more of this shit happening with people putting themselves at legal risk (or worse) to make this point, and it's going to be great watching who is appalled, who laughs and who smirks and nods.

- People who don't accept the tactic are hung up on the fact that it won't change any minds. So what, nothing does. Nobody changes anybody else's mind, ever. That's not a thing. People change their minds on their own. All anybody else can do is be right about something and say so.

- I think it's important to make correct observations and provide witness to monstrosities. Complacent people don't like the disruption because they value order more than anything, even the survival of humanity, and they can't abide anyone not indulging them in that preference.

- I appreciate their discomfort because it makes their priorities clear. You will hear a lot from people who have never visited or thought about this painting prior, how "ruining things is pointless." Yes, quite. Hose off the painting and it will be fine. Not so much the planet.

- I will concede that they wasted soup.[22]

CHAPTER 8

The Big Yes

WHAT WORLD DO WE LONG FOR, AND WHAT DO MOVEMENTS TO create it look like? In the words of organizer Sydney Azari, "I want future children to grow up in a world knowing that the people alive today loved them so much they were willing to risk everything so that they could live."[1] We know that means entirely transforming the ways that *we* live. We can move the needle on everything, from the energy sources that power our society to the goals and measures of our economy, from the ways we move through space to the way our food is produced, and yes, which foodstuffs we rely on as a society. We can end the fire sale on nature—we can reforest, rewild, and tend it. All of this also means changing how we treat each other.

The strategies to accomplish this goal are the Big Yes.

The Big Yes is invigorating, and unlike the Big No, Yes grows beyond the confines of the problem. Yes! Our future is brightened by as many good ideas as we can conceive. Yes to community networks of love and support. Yes to comprehensive, free healthcare that does right by pregnant people and youth. Yes to food systems that regenerate soil and culture rather than only extracting from it. Yes to reparative systems and anti-racist economies. Yes to clean energy, Yes to free, efficient, comfy public transportation systems, Yes to superb public education. Yes to welcoming our neighbors who are fleeing catastrophic events elsewhere. Yes to afford-able, hurricane-resistant, cool houses and cities. Yes to backyard gardens, to reforestation, to sustainable aquaculture. Yes to art everywhere. Yes to knowing our neighbors, yes to being able to retire with dignity. Yes to gender-inclusive parental leave. Yes to universal suffrage and voter access. Yes to restorative justice, Yes to community solar. Yes!

The Big Yes describes everything—that we've already imag-ined and that *we've yet to imagine*—that we can do to make our world climate-safe and socially just.

But in the face of so much potential and so much need, fig-uring out how to start can also be overwhelming. And so we have four guiding principles to help (these may look familiar). At its core, the Big Yes is:

1. Bigger than you.

2. Inspiring to you.

3. Powered by a vision of a life worth living.

4. Possible. That is: a project, not a general idea.

Scale is important to this definition. When we say that Yeses are bigger than you, we mean that they are political—they are *collective*. We do these things with other people, for all our bene-fit. And when we say that it's a project and not a general idea, we mean it's a step toward the vision, not the vision in its entirety. This is both a recognition of our individual needs and limits, and a reminder that the future is made collectively.

The Big Yes is nourishing. We each possess the knowl-edge of what makes a life worth living, in a community worth belonging to, on a planet that feels like home. We possess this knowledge, and it does us good to cultivate it. Life can't be only a slog—that won't sustain us. Sometimes this gets caught up in the commodified discourse of self-care, but we prefer adrienne maree brown's formulation of what she named *pleasure activism*:

> Pleasure is a feeling of happy satisfaction and enjoyment. Activism consists of efforts to promote, impede, or direct social, political, economic, or environmental reform or stasis with the desire to make improvements in society. Pleasure activism is the work we do to reclaim our whole, happy, and satisfiable selves from the impacts, delusions, and limitations of oppression and/ or supremacy. Pleasure activism asserts that we all need and deserve pleasure and that our social structures must reflect this. In this moment, we must prioritize the pleasure of those most impacted by oppression.[2]

In other words, the journey and the outcome are linked. Feeling good is a priority, and a compass, in the practice of doing good. Everybody knows that *doing* good doesn't always *feel* good, and sometimes what feels good is not good for us (or others). But we understand that joy is our right.

This is not to suggest that every moment of every Yes will be fun (spoiler: it won't be). The Big Yes sees our potential for transformation. But transforming a system is made up of a ton of smaller jobs, very few of which feel transformative in the moment. There is a lot of deeply unglamorous work involved with saying Yes: some days Yes looks like literally shoveling shit for the community garden, crunching numbers for solar installations, or making a monster pot of chili while collating handouts for a meeting. Some days it looks like sitting on an endless video call, being patient with your co-organizer's patchy internet. Getting out the vote can require hiking through unfamiliar neighborhoods and knocking on doors of people who may not want to talk with you; phone banking and text banking can be tedious. And thankfully, some days Yes looks like a dance party. We get to the big, visible, enlivening climate victories by way of a ton of less visible, less sexy work. Luckily we get to do it together, and we can take turns.

The Big Yes is not blind optimism. We've spent a lot of time debunking Micawberish optimism in this book, and for now we'll simply note that even if all of us say Yes at the top of our lungs today, we'll still have climate consequences to deal with tomorrow. On the occasion of the 2017 presidential inauguration, 350 Director May Boeve shared this sentiment: "In the coming years, we will need to hold onto a complex hope that never forgets the damage being done, and keeps sight of the huge number of people stepping up to fix it." Yes is not a substitute for No, for hard work, or for feeling our feelings. Rather, the Big Yes is a step we take together toward a world we want to live in.

In the winter of 2023, Meghan gave a talk at a climate research institute at Northeastern University. It was a dynamic group: a mix of undergrads, doctoral students, and postdocs working across the natural, physical, and social sciences—and interestingly, all were women except for the director. She asked members about their own Big Yes.

"I want to see the stars in Boston and New York City," a woman named Kourtney said, without a moment's hesitation. In Meghan's words, "her comment gave me the chills: the immediacy and the clarity of it. And flattened me in the best possible way—it sparked a kind of homesickness for a world that I'd never actually inhabited." She continues:

> That Yes awakened something in my own imagination as well. And it offered direction toward the conceivable future, more broadly. I eventually recovered my powers of speech, and we talked about how that vision of stars in the city could illuminate the steps to get there: looking at light pollution, which in turn led us into a conversation about light emissions, which in turn led us into a conversation about city planning, and the kind of dense, walkable neighborhoods and plentiful parks we would need to build to cut down on light pollution in areas where people live. Glimpsing—even momentarily—her dream

of dark cities beneath twinkling stars put an unexpected quiet into my body.

So why do we struggle with saying—and imagining—yes?

So if the Big Yes is so amazing and magical, why don't we already live in an eco-paradise? We think there are several reasons why we struggle with the Big Yes. You may have picked up on one theme already: the way the climate crisis is framed in the mainstream inhibits creative expression—it's deadening. When we focus on our carbon footprint, we inhibit our ability to be creative—and *big*—about solutions. There is no meaningful, inspiring solution that can be enacted just at the individual scale.

For some of us, the seriousness of the problem, and the limited emotional vocabulary we've developed for climate, makes any talk of joy, beauty, or fun, seem impossible, or inappropriate. People are dying! A focus on Yes can seem frivolous, in poor taste. Even if we've made some space for the diversity of feelings in our own hearts and personal lives, we may feel pressure to don the mourning clothes when it's time to take action. In their book *Joyful Militancy*, carla bergman and Nick Montgomery challenge the idea that good mood means bad politics. For them, "the point is not that happiness is bad or that being happy means being complicit with empire. Happiness can also be subversive and dangerous, as part of a process through which one becomes more alive and capable . . . joy, in contrast to happiness, is to move away from conditioned habits, reactions, and emotions. . . . It is a process of coming alive."[3]

Also, we are surrounded by harm. We wrote this whole book about the harm to bodies, families, communities, the land and air, and what meaning we can make by recognizing the full situation we're in. We are calling for acts of creativity, but seeing harm, describing harm, and working to stop harm, takes a big toll. It is hard to create—or feel creative—when we're tired, afraid, or in pain. What we're able to imagine may be limited by what we've experienced, and what we've come to believe is impossible.

Whereas some of us struggle with *saying* the Big No, we struggle with just *imagining* the Big Yes. Being able to say No is a prerequisite for being able to imagine a real Yes: you have to understand a problem if you mean to fix it. You have to understand a system if you mean to replace it. But being able to say No does not necessarily mean you have the time or emotional energy to dream big. In one way the Big Yes asks for more of us than No, because it needs our vulnerability as well as our strength. Real solutions arise from our desires, and our connections to each other. And just like in intimate relationships, to share our desires is to risk disappointment, judgment, rejection, failure. The work of deepening our connections is full of this kind of intimate risk. We are learning how to be open and connected, to develop solutions that are bigger than the problems they fix.

How do we learn to do it better? Practice!

Meghan describes the social practices that are now a part of her band practice:

> One of the greatest joys in my life is playing the trumpet in a band of twenty-five or so people. We're an activist street band, which means that our main goal is to make the experience of being alive a bit more fun and weird. We believe music changes things. And so we show up for protests and rallies and celebrations and fundraisers and parades, and festivals all over the country. My band has always felt to me a little bit like a portal into the future: where activism is forcefully joyous, where imagination is key, where saying yes (and saying no) can be a party, in addition to being deadly serious. We've played at massive political demonstrations like the Women's March in 2017, trying to be a heartbeat among millions of anguished people; a doctor snuck us into a hospital via freight elevator for a surprise performance for patients; security chased us out of a mall as we rode up and down the escalators playing the Tetris theme song

to the delight of shoppers and the chagrin of management; one summer on the way back from a festival, when traffic stopped on Interstate 90 in upstate New York due to an accident, we piled out of our vans, climbed up onto the guardrails and played an impromptu (barefoot, hungover, but enthusiastic) show to several hundred people stranded in their cars. We've played children's Halloween parades in our own state, and Mardi Gras in New Orleans.

Mostly, though, the work looks like practice—practice in which we learn and perfect the music, practice that prepares us for these big inflection points. We practice together every week in a public park. We love each other hard and bicker frequently and keep an eye on each other's kids and dogs, and try to become better musicians and hold ourselves to account as people. We're activists, but it's not just activism that heals: we try to create freedom to feel within our own loving, chaotic space. The magic comes from belonging to a group that loves people and music equally, and that has big but accomplishable goals.

Who says yes?

Everyone has the right to imagine, name, and work toward a beautiful future. But we also want to note that there are patterns, in practical terms, in who *gets* to focus on their positive vision. A lot of people are in one way or another living under assault (through a combination of gender, race, economic, ethnic, and other oppressions), and by necessity focus their work on *stopping harm*. No, you may not regulate my body, no you may not dump toxic sludge near my children's schools, no, you may not turn my neighborhood into a sacrifice zone. And so we have to spend a lot of time and energy fighting off bad stuff rather than imagining the good.

On the flipside, privilege of any kind can have an especially perverse flavor in climate organizing. Some of us may be prone to discrediting ourselves—or others—from taking action because we feel we don't have a direct claim to the present trauma of it. Perhaps you pack away concerns in a box labeled "First World Problems," or, because you may not be a member of a frontline community, you may feel shy or imposter-ish about participating. We've seen privileged would-be activists use a faux acknowledgment of the harm that others experience as a reason to curl into themselves instead of getting active. Plenty of slippery cognitive things happen when people are thinking—instead of doing—about the climate crisis. We all know by now that our minds don't want to linger there: climate paralysis is very uncomfortable.

But the Big Yes vision arises from everyone's contributions. We can't let a fear of getting something wrong prevent us from doing what we know is right. A useful reminder for those of us with any sort of privilege is that the goal is not to rid yourself of that privilege, but to get working so that *everyone else* can enjoy the same social benefits you do. Having a big, beautiful dream doesn't require having a big, rigid ego. When we work humbly toward solutions with justice at their core, we usually find that our efforts are welcome.

WHAT SKILLS DO WE NEED TO SAY YES?

Many skills that we require to say Yes are the same we use to say No. And the two most important are inclusion and collaborative imagination—we need to be able to talk with each other about the world we want to live in. That means we practice sharing space, power, love, and accountability in a big collective. We note here that there are many excellent books that go into greater detail on community organizing (our favorites are in the appendix). Here we are sharing what we consider the essentials, to get the party started.

Extending welcome

Welcome people who come through the door. Extending welcome can mean creating small but crucial practices, like having everyone at a meeting introduce themself with more than a sentence so that you all learn a bit about who is in a room together. It can mean making a point of saying hello to the person who showed up by themself. It can literally mean having enough seats—and big enough seats—at a community event so that nobody is left leaning awkwardly against the wall. There are lots of ways to include more people in a get-together: childcare and schedule considerations for parents and 9-to-5ers, interpreters for your local languages, elevator and ramp access if you're up any stairs. And in our experience, food is friendly: any community meeting goes over easier with pizza.

Welcoming people is not just behaving kindly, but also having useful things for them to do. As Jen Mendoza told us in the last chapter, a major reason that people don't stick around a campaign or project is when they've shown up and there's no job for them. For organizers this can mean waiting to invite volunteers until your core group has figured out practical steps you need help with, or being clear in your invitation that the day's task is to brainstorm those practical steps. Welcome can also look like a varied list of jobs that people get to choose from.

Juan-Pablo Velez, with whom we first spoke in chapter 5, is the founder of WinClimate, a think tank that, in its own words, "uses data science to support legislators and advocates working to pass state-level climate policies." He brings a scientific enthusiasm to his organizing efforts. When asked why he thought more people don't get involved in activism, he told us flatly:

> The user experience of activism sucks. In activism, you're trying to get somebody to take an action. Marketing has the same goal, right? You're just trying to get them to consume something. What strikes me is the massive gap between the sophistication, and customer-centricity of marketing, and the [lack of the] equivalent thing in activism . . . What I see over and over is just this assumption that the people we're trying to [reach] are just like us. I think you should be maniacally concerned with,

who are the people you're trying to reach? And what does it actually take to get them to do a thing, particularly when, at first glance, it's something that doesn't have the dopamine rush of buying something? It seems like a chore, not like a treat. So it's that much more important to figure out exactly who it is you're trying to get. Experiment. Talk to them the way they actually are, and not the way that *you* think about the world.

While not every scrappy little group has the capacity to crunch numbers about their potential recruits, the point here remains: by creating spaces where *people actually want to be*, they're more likely to show up, and if you talk with them about what they're good at and what they want out of their involvement, they're more likely to stay.

Listening well

Listening is one of the most fruitful skills to cultivate as organizers, neighbors, community and family members, and just as people sharing the planet with others. In his paper, "The Believing Game," writer and teacher Peter Elbow describes how our culture has undervalued empathetic listening—*"to dwell genuinely in ideas alien from their own"*—or indeed misrecognized it as a social gift, rather than an intellectual skill that can be learned, taught, and practiced:

> When we see [people] listening and drawing out others, we call them generous or nice rather than smart. We don't connect good listening to intelligence. We say "Isn't it wonderful how they can mobilize others and actually get things done," but we see that as a social and personal gift rather than an intellectual skill. We call creativity a mystery. And because our intellectual model is flawed in these ways, we don't teach this ability to enter into alien ideas.[4]

To listen, we have to step off our own merry-go-round for long enough to hear and absorb the experiences of others, and let our own minds digest unfamiliar ideas. We can *plan* to listen: other people's views require actual time and space on the agenda. When we are listening, our nonverbal behaviors are just as important as our verbal ones. To show our conversation partners that we're actually tuned in, we use focused, nonthreatening body language. Phones and other tasks are put away. Arms are uncrossed. Eye contact, nods, and subverbal listening noises may help.

Asking yes-or-no questions will produce yes-or-no answers, which can end a conversation rather than beginning it. Asking open-ended questions that are free of expectations or assumptions may elicit answers you haven't predicted. Examples we use when active listening include:

- Can you tell me more about that?
- How has this felt for you?
- What came up for you when that happened?

Patience is the core of listening. We can listen to understand rather than to respond, and we can let ourselves be affected by what we hear. We can allow other people to speak without interruption and at a pace that makes sense to them. Even when it's uncomfortable, we try not to prepare a reply while the other person is still speaking, or rush to fill periods of silence with our own thoughts or stories. If any of this sounds so basic it's silly, just remember the last bad conversation you had. If it wasn't today, it was probably yesterday. We can all use practice.

Those of us with privilege within our identities particularly—whether that privilege is gender, racial, class-based, religious, or any combination—can avoid doing a lot of unintended harm when we're listening well and often. Saviorism is when I believe I already know what's best for you. This form of not-listening is widespread and well-documented in activism. Juliste tells us of the challenges she has experienced organizing with middle-class communities in Chicago. She says of that environment, "it's very macho, but you can't tell them they're macho because they think they're good feminists. Which is harder to deal with, honestly." *There's* her eye-roll. "I've experienced more transphobia working for some of the big and medium greens [NGOs] than I do on the construction crews I normally work on."

She continues, "I'll never forget some organizers, like with Greenpeace for example, saying to me and people in frontline groups, 'oh, you know, we bring a certain professionalism to activism, and you're not going to be heard without us,' or 'you need this, you need that, and we're here to offer it for you.' And they don't actually think of what building a community *is* and what building a citywide coalition of groups is." She pauses. "Any injustice happening, there's going to be some kind of toxic goop being dumped on some neighborhood, smog-filled, or like toxic dust all over, or incredibly lead- and mercury-filled soil. And they just don't speak the language. They could *literally* not speak the language. . . . And like, three toxic white dudes can kill a whole citywide movement. I've seen it happen over and over."

To open up rather than shut down possibilities—and movements—we can do as Peter Elbow suggests and honor good listening as intelligence.

Building relationships

We've spoken elsewhere about how a group becomes a community. Coming together takes effort, and staying together requires practice. Rabbi Shoshana, whom we met earlier, describes her journey into community organizing:

> I somehow had the idea when I was little that I could—and should—save the world single-handedly. And I didn't totally not-believe that until I graduated college and completed a

year-long fellowship where I learned the basics of community organizing: that relationships are the building blocks of power, that those people can then come together to effect change through power mapping and strategic action. It was revolutionary to me. And it wasn't just about the outer change. I also learned that I could feel *sustained* by those relationships. . . . I learned that relationships can be thick and beautiful, and collectively buoyant, a raft where we're all holding hands, and the raft can carry us above the dangerous currents.

The Big Yes is a collaborative act and in community organizing, social skills and enjoyment are obviously helpful. Jen Mendoza tells us, "I'm a people person. I love just getting on the phone, having a one-on-one with someone, and inviting them to the meeting. I'm a campaigner but I'm an organizer first, like, through passion of people and being genuinely excited about building out the movement." But even if you're not naturally gregarious, you can be an excellent organizer. And organizing skills—while they take a lifetime to master—can be acquired in small doses. No know-how is too small to be put to good use.

If you're feeling stuck, there are many groups that teach basic community organizing skills. We recommend Movement Generation, a San Francisco–based group that hosts political education programs ranging from in-person training and multi-day retreats to online courses. Movement Generation's stated goal is to build shared understandings of the impacts of the global ecological crisis and ignite strategies for a just transition. They're a favorite of ours because of their climate justice framework, but there are many other groups out there, and any class can be a jump start when we're stalled out.

One of the most powerful organizing skills is the ability to build bridges between different groups of people in order to accomplish a shared goal. Does your group want all the buildings in your city to be bound by a very green building code? The first step is identifying others who would be interested in that outcome. In addition to getting environmental groups on board, you'll potentially need architects and builders, as well as transit advocates, anti-poverty activists, and housing organizers. Next step is to bring them together.

As a policymaker, Meghan believes the trick is to be led by the vision rather than purity tests, while remaining true to our own values and principles. It's a funny line. She says,

> I'll work with people whose attitudes I really dislike if it means that we get climate legislation across the finish line. Politics is how we resolve our differences without violence, and that does not mean that we'll love everyone we're working with all the time. That's not what it *should* mean. Purity tests largely do not serve me—if I need someone's vote on a bill and I think I can get it, I'll speak to them about it. I was elected to pass legislation, and while I may be both personally and politically offended by someone's view that I'm not deserving of the same autonomy over my body that he is, for instance, I do not feel that it dirties my hands to enlist the support of an anti-abortion colleague for a climate-related bill. Nor does his support for a climate bill lessen my commitment to reproductive protections.

Like most worthwhile projects, climate organizing is a long game. All the activists we spoke with underscored the difference between an action and a movement: "If you and I ran out and blocked Diversey Avenue over there," says Juliste, pointing across her neighbors' backyards, "we could do it. We could stop traffic and we would say, you know, 'End Climate Change Now!' I don't think it would amount to anything. We could get two hundred people to do that right now. But we don't have anything backing it up. This stuff is long haul."

Jayeesha has seen a lot of one-and-done activism, which she thinks "comes from flash-in-the-pan issues or events, or people just wanting to be part of the cool new thing." But this isn't relevant to movement building. "That, to me, isn't anything that sticks.

I *do* think some people get a lot out of this, the very shiny, very media-fueled, hot button movement moments. Some people are then activated for a long time. It does build, but I think it's probably not the way that most lifetime activists are birthed or created." We may enjoy the flashbulb moments of success, but those moments are built slow and steady with long-term commitment toward a long-term vision.

Imagining better

We began to unburden our imaginations in the last chapter, removing inhibition from our chests to say the Big No. The Big Yes calls for more. American institutions are not typically imaginative; rather, they trail the public imagination. If the Big No is about undoing the bad, the Big Yes is about creating the good, often from scratch, often equipped just with the designs of what we *don't* want. These are complex undertakings, and it is difficult to find the space, quiet, and time (and sleep!) typically associated with creative expression amid the clanging alarms and shouting matches.

Jayeesha tells us that fearmongering, and a lack of imagination are the biggest flaws she sees in the climate movement. "I feel," she says, "like a lot of the climate movement operates from the vantage point of agitating. Agitating from a place of fear, and anger. It's even in a lot of the OG organizing stuff, right? Agitate! Agitate! People will be moved if they're angry or fearful!"

Her expression is weary. "And I just don't think that's going to work. That's especially not going to work for the climate crisis—it could work for the labor movement, because people get pissed off about their jobs. But if you get really pissed off and scared about the climate crisis, it's not the same kind of thing, because it's just gonna feel so beyond you. [Then] it's much easier to shut down, right?"

The climate crisis casts its inhuman shadow. Everyone alive today will be living with its consequences for the rest of our lives, and the adrenaline of anger will not sustain us to transform societies. Jayeesha continues, "so to me, the way to really organize and build momentum in the climate movement has to be through the creation of vision, possibility, solutions. A just and collaborative way of being, instead of a fear-driven competitive mentality. Colonialism, capitalism, and white supremacy, everything that got us here; what can get us out is *not* those things, right? It's the *antidote* to all of those things."

Much of the climate movement continues to deploy a communications strategy of fear, anger, and panic-inducing statistics. But for Jayeesha, one of the ways that we can flex our creative muscles is to look to people for whom imagination is *already* how they do things. For her, the skill that is most underappreciated is art-making: "It's artists and cultural work. Like art, art, art! Culture, healing. It's something that [campaigns] add on, or don't have at all. Especially the more 'radical' and the more activist and the more 'badass' you get, the less you'll see those kinds of needs being taken seriously." She continues,

> That's why I think the climate movement is not doing great right now, is because it's become such an angry, fearful, data-driven, like, numerically-oriented movement, as opposed to being relational, and caring and really modeling the world we want to see as the way that we build.

In the words of scholar, writer, and activist Toni Cade Bambara, "the role of the artist is to make the revolution irresistible." As Jayeesha points out, the path and the destination are linked. One creative exercise is to think about the parts of our lives that make us feel good, and how they can connect to aspects of the world we want to inhabit. When we put energy into a creative practice we're flexing our imagination muscles, increasing our potential for pleasure, and exploring possibilities. For instance, Josephine is an illustrator and Meghan is a musician; both of us have used our art as a way to generate pleasure, develop our thinking and problem-solve, as well as to explore ideas with others. In fact, our theory of the Big No and Big Yes emerged from one of Josephine's sketches during a Conceivable Future discussion.

An important note here about the creative process. A lot of us who make artwork know that the kryptonite of brainstorming is editing. Your imaginative practice is a personal exploration of a future that gets *you* excited. It starts as a rough draft, a sketch, a freewrite, a blind contour drawing, a melody you hum with nonsense words into a voice memo. It's intimate, can be private, doesn't have to be politically or aesthetically perfect from the start.

Once you get familiar with your fantasy, and you gain some practice expressing it in different ways, *that's* the vision you'll share with others to create your collective goals. We use creative methods to attain creative goals. So to begin this practice we invite you to take a minute, close your eyes, and think about a world twenty years into the future that you inhabit. You don't need to pretend it's perfect, but imagine how you can live creatively, healthily, and joyfully. Notice:

- What it looks like. Are you inside, outside? In a city? In a village? Out in the countryside?

- What does it sound like? Can you hear animals, people, trains, rustling trees? Are you energized by the bustle of a large community or sustained by silence?

- What does it smell like?

- How do you spend your days?

- How do you move around? Are you walking, taking the train, biking, or in an electric car?

- What are you eating? What does it taste like? Where did you get it?

- Who is with you? Who loves you? Who do you love? Who is responsible for you, and who are you responsible for?

This list—although a little facile—is intended to take some blank-page pressure off the imaginative exercise. Looking inward to tend *your own vision of beauty* is a solid place to start, and can give you a sense of what you want us all to grow toward, together.

SIDE EFFECTS OF THE BIG YES

So what are the downsides to the upside of climate action? As we said before, the Big Yes requires our vulnerability as well as our strength. And this makes us—surprise—vulnerable.

- We may be vulnerable to self-doubt, worried that our Yes isn't big enough, or that it won't succeed. And there are no more guarantees with Yes than there are with No: Big Yes projects benefit from failure plans as much as their No counterparts.

- Big Yes activists may also be unwilling to take strategies of No seriously, which can cut us off from many excellent colleagues.

- A belief that Yes is the only valid strategy can even cut us off from reality, keeping us in a kind of ungrounded optimism.

- Yes work may make us feel like we need to be "up" or happy all the time, which is exhausting, and can create an undertow of negative emotions.

- Yes *organizations* may not make space for a full range of feelings, which can be painful, alienating, and even hasten burnout. Burnout can be a surprise because Yes work seems like it shouldn't be the heavy one.

The Big Yes can be vulnerable in another way: big, bold ideas attract trolls. People behaving sincerely in public often become lightning rods for hate. When there's no flaw in your argument and nothing wrong with what you're doing except that it threatens someone's profits or fragile ego, ad hominem attacks are all that's left. That can take the form of violent threats and sometimes even literal violence. And it's on the rise as a form of political activism on the right. We've found that ignoring, blocking, and reporting trolls is usually the best strategy. Sometimes, as in the case of a smear campaign against Meghan and a colleague's psilocybin decriminalization bill, the trolling attempt actually raised the bill's profile and generated public support for it. But being the target of bullying is never fun, and doesn't necessarily go well, and over time it takes a toll. Be ready to spend some amount of energy on managing this crap.

We bring these downsides up as another reminder that we need to stay grounded in the bigger picture: the movement is bigger than us, bigger than our campaigns, organizations, goals, and personal ideals. Big Yes and Big No can have each others' backs. As you go to bed exhausted, someone else is waking up, jazzed and inspired and ready to do their part. And if you find one day you can't extol the virtues of your Yes solution one more time, there's a worthy No campaign right around the corner, guaranteed. There is always another job to do, and another way forward, when you're ready for it.

VARIETIES OF BIG YES

For the rest of this chapter we talk with people who are doing Big Yes work, at different intervention points of our society. The same framework that served us for No also serves us for Yes—we've reprised our chart to reveal some order in the complex ecosystem of climate solutions. These profiles are illustrations and inspirations, not prescriptions. We're not offering everything happening at every level, but rather some imaginative courses of action. We want to help you see yourself in the Big Yes.

And it bears repeating that although our goal is to change the systems that govern our lives by finding the biggest levers we can reach, there is no single lever—no magic bullet—that does it all. Every system, whether it's a supply chain, a social norm, or an HVAC, consists of many component parts. The good news here is that you can choose where to start change-making from lots of different places; it will take lots of different kinds of work.

The bigger a system gets the more harm it can do, but the more abstract it becomes. Abstract systems are the hardest kind for most of us to imagine changing (e.g., racial capitalism or sexism in healthcare), and that can feel paralyzing. This paradox of scale—e.g., the root cause of the climate crisis is centuries of global capitalism and extractivism, but we experience it right now at the level of self and community—often leaves us sitting defeated in the bleachers when of course we'd rather play.

The Big Yes means turning "change the system" from a vague piece of rhetoric into specific, nameable, *doable* steps. We don't love concrete, but we do love concretizing abstractions.

At each of these scales, the Big Yes is primarily concerned with changing systems—from tiny local ones all the way up to large national ones. These systems help us manage life, experienced in the form of rituals (e.g., marriage), traditions, customs, institutions, and laws. We bake our values into these aspects of everyday life so that we don't have to think so hard about everything all the time. The ubiquitous, unconscious nature of these systems makes them easy to take for granted, and hard to change.

Hard as it is, creating better systems—making our institutions and practices more just, nurturing, and loving—turns good ways of doing things into our default setting. For instance, if we require that every new building in the state of Connecticut is hurricane-resistant and energy-efficient, then that standard is the *baseline*, and we no longer have to argue about efficiency

	Municipal or Local	State or Regional	National
Socio-cultural	• Mutual aid • Bike jams and transit parties • Freestore and tool libraries • Disaster preparedness	• Campaign for Indigenous People's Day • Creating reuse markets • Resilience plans • Festivals and fairs	• Media campaigns • Public education campaigns • Creating uses for recyclable materials • Resilience programs and funding • Creating new holidays
Political	• Elect great candidates to local office • Referenda • Community solar/wind • Municipal composting • Zoning for farming (urban and rural) • Zoning for transit, density and walkability	• Elect great candidates to state office • Budget • Rights of Mother Earth Legislation • Watershed protections • Defend treaty rights • Regional transit systems • Codify state abortion rights	• Elect great candidates to federal office • Efficiency standards and subsidies • Funding climate-connected issues (e.g. hurricane-resistant affordable housing) • Single-payer health insurance • Insurance coverage for preventative healthcare (e.g. doulas)
Infrastructural	• Public transit • Community gardens and farms • Bike lanes and walking paths • Accessible healthcare/clinics • Green roofs • Public charging Stations • Green building codes • Incentivizing good solar siting	• Net metering • Stormwater • Wastewater • Renewable energy production (and jobs!) • Public transit • Efficiency standards • Wildlife corridors • Incentivizing good solar siting	• Rail expansion • Renewable energy production (and jobs!) • Interstate charging station network • Shoreline rehab and protections • Resilient smart power grids • Reproductive support (e.g. Momnibus)
Economic	• Co-ops (for all kinds of businesses!) • Local food procurement • Fundraising! • Electrification/weatherization incentives	• Regional food infrastructure and procurement • Credit Unions • B-corp certification • Green pensions • Medicaid coverage for reproductive healthcare	• Green, ethical financial services • Universal basic income • Just transition programs and funding • Disaster relief fund

Figure 8.1

requirements for every individual project. This saves not just heating fuel, rebuilding costs, and heartache, but advocacy effort, which can now be turned elsewhere. If we win workplace policies like parental leave (rather than only maternity leave) then we normalize expectations that fathers and parents of all genders have a right to the full privileges and responsibilities of parenthood. When we require changing tables in every bathroom, we transmit the same message. These are legislative examples of how good institutions can raise the baseline of social expectations.

The same approach can work in more informal situations, too. For instance, Meghan's band observed an ongoing issue with gender representation, and they decided to tackle it at the system-level. They found that during breaks in the music, it was mostly men who were stepping up to take improvised solos. In their group, men were feeling confident to walk into the literal spotlight for a minute or two and they would do so quickly, leaving little space for anyone else. The group noticed the pattern and determined it was a problem, resolving to cultivate confidence in band members of all genders.

So now, before every performance they choose the soloists preemptively, making sure that their gender representation is more or less equitable throughout the course of a show. The men in the band have had to practice stepping back and making space; everyone else has practiced stepping up. And while no one is *required* to solo, by shining a light on what used to be considered an "automatic" or "natural" process (and creating a very intentional alternative) the group has fundamentally shifted the distribution of who takes solos. In doing so, they have also interrupted a bad-news gender dynamic and replaced it with something more open and welcoming, cultivating a crop of inspired new soloists. The process mattered: it wasn't just the path to their goal, it *was* the goal.

BIG-P POLITICS

One of the biggest levers we have at our disposal is big-P Politics—by which we mean formal democratic politics, and voting. We sometimes hear of activist strategies being divided into "inside" and "outside" strategies, meaning that they do or don't formally engage with the democratic process to make change.

This can be a helpful distinction in some cases. But all of the activism we're talking about here—ranging from issue campaigns to civil disobedience to solar farms—happens in the context of the political life of this country. In other words, there's no such thing as being wholly "community-based" or wholly "political" in one's activism, all outside or inside—they are entangled. Every type of engagement that we describe here both contributes to the political reality of this country, and is shaped by it, however subtly. A composting co-op is governed by municipal waste laws; the solar farm is subject to zoning; public transit is funded by the state. Whether or not you "do politics," politics do you.

Meghan feels strongly about participation:

> I understand that politics can be alienating. They can be alienating for me, and I'm right in the middle of them! And there are good reasons for this—by serious measures, the national political scene in the United States isn't even a democracy anymore, but rather an oligarchy.[5] Also we've defunded civics education for the past three decades, so there's not a ton of literacy about political processes, and a lot of people are having to teach themselves how to get involved. Voter disenfranchisement is real, and not everyone who is eligible to vote can do so practically. But even with all of those things in play, I believe that policy is a place where big levers are reachable, and that's why I'm in politics. I was tired of my vocabulary being only the Big No. I wanted to say Yes at a bigger scale with others. For me, politics is a way to do that.

> I also believe that, in the United States, the state can be an especially effective level for climate legislation, particularly given the gridlocked mess that is national politics. State and regional organizing usually feels less daunting than national campaigns, but can move the needle for huge numbers of people. Our State Capitals are geographically closer and more inviting than Washington DC for most of us, and our state and local representatives tend to be much more accessible than

our national ones, simply because there are more of them and they usually live in the community they serve. And this is a level where meaningful policy both is—and importantly, also feels—achievable.

During my second year in the Rhode Island Senate, we passed our Act on Climate bill, which created legally binding emissions targets for the state. There are fewer than 2 million Rhode Islanders, and creating and instituting that was a meetable challenge. In 2022 Massachusetts passed the Millionaire's Tax by popular referendum, aiming at a more equitable tax code to generate money for public transit. In the same year, Kansas codified the right to abortion the same way. Those wins were accomplished by people doing a lot of grunt work: knocking on doors, having conversations, turning out voters. US states are medium-sized political units: they are big enough to enshrine real climate commitments and real protections, and small enough to organize. Politics has a reputation for being "dirty," but a state protecting its right to abortion access by referendum is a big damn deal. And it was accomplished with people power.

Politics are how we resolve our differences without fighting and violence. Politics can *feel* dirty—negotiation often does!—but it's certainly less dirty than war. Politics involves disagreement and sometimes it comes down to choices between shit options and shittier options. As with most of life, there often aren't wholly right or wrong answers.

But opting out is not a noble choice; it doesn't let anyone keep their hands clean. And when we can set aside our cynicism long enough to do the work, there are outsized wins being had all over the place. We'd rather have a Massachusetts with a Millionaires' Tax directing money to public transit than a Massachusetts without one! Absolving ourselves of our civic responsibilities just lets someone else make bad choices in our name.

To that end, we see voting as a part of a civic practice. It is not, in our view, activism, and it is not enough, unto itself, to solve any problem. Voter *mobilization* is activism and supporting voter *access* is activism. We'd equally extol the virtues of flossing your teeth and voting; we hope you do both if you possibly can.

The first step to making political change is to build power. When Juan-Pablo Velez figured out how to actually do this, he got passionate about political organizing. "What really changed my outlook was understanding how you win in politics, and the fact that you *can* win in politics," he tells us. "Because I had this fairly typical idea of politics as being: you have these elections. You elect a Congress, you elect a president, and then you cross your fingers."

This shifted for Juan-Pablo when he began to get involved in city-level organizing. He observed as a political campaign for congestion pricing emerged. "I knew how to collaborate with people in government who were trying to do good things. But I was scared to be confrontational. And moreover, I didn't know that *there was a method for winning*." He describes meeting the founder of the Riders' Alliance, a group that had successfully pushed for congestion pricing in New York. "And he just told me how they did it!" he says. In New York,

> they were able to essentially push [former NY governor Andrew] Cuomo to do it with a tiny group of people, just by using the right strategy. And that strategy is pretty straightforward. It's like: politicians want to get reelected, even the good ones. So the way you can build power over them is to take away their votes, or their good press, or their money. And the way you do that is by organizing people and manipulating the media. And there's a whole codified playbook for how you do this. And the more you look at what progressive laws *do* get passed in cities and states, almost all of them come from some campaign run by some advocates. Like, it's just how the whole world works!

Understanding a set of identifiable and clear-cut strategies was a game changer for Juan-Pablo. "No one had ever painted it that way for me," he says. "And then it turned out like, oh, no, the way anything ever happens is by building power and influence over politicians. *And it's not accidental.* It's methodical. And it works!"

He still looks a little stunned by this information as he concludes, "that completely blew my mind. For me it was just like stumbling into the secret of how you organize to win any political policy. And I feel I'm on this mission, like, people need to know this. People need to know that this is not just viable, it's how *things actually get done.*"

These playbooks exist—political organizing is a teachable and learnable skill set. And even within our highly imperfect democracy, it can be one of the most effective ways to say Big Yes, at multiple levels.

LOCAL

The chart we offered a few pages ago is a way to visualize our to-do list. The least abstract way to plug in will be at the local level—in our cities or towns, or even our neighborhoods. And if you're stuck there, the number one recommendation we have? Read your *local* newspaper. City, town, and state coverage is where the good, doable, joinable stuff gets coverage. Here are a few initiatives where people are working to build the world they want.

Better than Recycling

Some Big Yeses are based on a simple vision of being less stupidly wasteful. One of the most common laments we all hear is about the sad state of recycling in the United States. Specifically, most of what we recycle doesn't really get recycled. And how about our consumerist focus on recycling instead of those other two R's, reusing and reducing? Chicago notoriously has the nation's worst rate of city recyclables actually being recycled: 9 percent in 2018.[6] And while there are many efforts within the city to improve the recycling system, several grassroots projects have emerged to take materials directly out of the landfill pipeline.

The first is the Waste Shed, which sells art supplies that have been donated or diverted from the trash. Founded by Eleanor Ray in 2014 in a back room of an art space, it formalized in 2015, has moved into larger and larger storefronts, and opened a second location in Evanston in 2022. The Waste Shed, beloved by local parents and other creative people, has become an essential resource to Chicagoland teachers, who can take materials for free for their classrooms. In 2022, the organization gave new life to 61,218 pounds of materials intended for the landfill. And their vision is about more than materials. Their mission statement declares that "while sustainability is the ultimate goal, creative reuse foregrounds practical, economical, and fun hands-on activities as points of entry into larger discussions about the things we own, and the way we live." Yes!

The second is a wonderfully scrappy operation called Eco-Ship Chicago, which collects clean packing and shipping materials—much of it plastic—in good condition for reuse. Founded by Aleksandra Plewa, a small crew of volunteers operate out of a storage locker, hold monthly collection events, and make materials available free for pickup to companies and individuals. Money donations are obviously also welcome, but their central ambition is to combat waste and promote sustainable business practices. They want "to connect local communities and businesses in a way that is not only beneficial for all, but also healthy for our planet."

Neither the Waste Shed nor Ecoship Chicago have ever struggled for lack of materials, sometimes even halting drop-offs while they catch up on processing the donations they've already received. But both groups have policies against "wish-cycling"—that is, donations of unusable materials in the hopes that someone else can figure out how to "do something with it." Anyone that starts to divert goods from the waste stream learns quickly how massive that stream is. The scale can be daunting, but doing—rather than wishing—is the process by which we get realistic about what we *can* do, instead of hovering anxiously between hope and the trash. Wish-cycling, as counterproductive as it is, comes from a good instinct: most people hate waste, and don't want to participate in wasting things. The success of these reuse enterprises points to a widespread desire for better ways to reuse and reduce, rather than just "recycle."

Fare-Free Buses

We feel the absence of good public transit keenly in cities all over this country. And we can get bummed out by the huge investments needed—and lacking—to build new (or maintain the old, sorry NYC!) subway lines and the high-tech, high-speed transit options enjoyed in other parts of the world. But one beautiful solution to our inadequate public transit requires no new infrastructure or technology, just political will and funding.

It's the free bus.

In December 2022 the city of Washington, DC, announced that the city buses were going to become fare-free, permanently. The COVID-19 pandemic laid bare for the District of Columbia—and cities across the country—that public transit was a lifeline for essential workers and that even modest fares could be a burden to them. During the height of the pandemic, cities like Los Angeles and Kansas City, Missouri, suspended fare collection to minimize human contact and ensure that residents with no other travel options could reach jobs and hospitals, grocery stores and offices.[7]

But increasingly, cities and states are piloting no-fare public transit as a way to do several things: increase ridership, save money for people who are lower income, improve the quality of transit through faster service, and reduce cars on the road. As of this writing, Boston is experimenting with several fare-free lines, and Meghan and her colleagues are piloting a free-fare program in Rhode Island's statewide bus system.

Free-fare arrangements are first and foremost an equity issue: people with lower incomes spend a higher percentage on transportation (up to 29 percent of household spending), and reducing transit costs benefits them first. Data from pilots across the country show that free fares can boost rider satisfaction, and that the buses work best when there is also frequent and reliable service. The faster boarding times have also been a big hit.[8]

Riding free buses doesn't get old. It's like a carnival ride—except it's free, and it goes somewhere. Meghan takes a selfie of her own shit-eating grin every time she boards the free route, and puts it on Instagram. Like Mei and Satsuki catching the Cat Bus in *My Neighbor Totoro*, when you catch a free bus you get to be the kind of VIP for whom a vehicle just magically appears and delivers you where you need to go.

This elegant solution may not be on your city councilor or alderperson's radar yet, or they may need more public support to make legislation happen. Advocating for buses can look a lot of different ways. Even a gas-powered bus is a big improvement on the equivalent bunch of cars, but cities buy new buses all the time: is there pressure to make their fleet electric yet? There may be a transit advocacy group already in your community. What are they working on? Improving bus service and policy is the fastest, most economical way to see a citywide shift away from cars toward the public good.

Community Solar

Juan-Pablo, whose political strategy epiphany we chronicled earlier, is a member of Friends of Columbia Solar, a grassroots solar advocacy group in Copake, New York, where he lives. Like everyone we spoke with, Juan-Pablo's journey to climate activism has been personal. As a young person, he was in Washington, DC, on an internship when a cap-and-trade bill was introduced that, in his words, "eventually went down in flames." He thought "Oh, wow, not only are we nowhere close to acting, we're not even close to having half the population acknowledge it's real!" He describes being "really freaked out by that difference in scale of the problem compared to the affordances of politics, but also getting kinda depressed and just sort of pushing it out of mind for about a decade." He went into informed denial: that kind of panic is a hard place to live every day.

> And that's especially true when you have detailed knowledge of what large scale action would involve. Knowing how important it was to have massive action right there and then, and knowing full well that it slipped away for probably another decade, [made] it really hard to be like, "I'm gonna keep thinking about it," because it just seemed like "what's the point?" I guess I didn't have the appetite for continuing to plug away at something when there was no opportunity for federal policy.

Disillusionment with the federal stalemate eventually redirected him to local activism. Juan-Pablo moved to the Hudson Valley during the COVID-19 lockdown of 2020, and there he happened into the local struggle over a proposed solar farm, after a chance encounter at a farmers' market:

> There's these folks handing out pamphlets and asking, "Have you heard about the new solar farm that's coming to town?" And I was like, "That's great!" They were like "no, it's terrible, it's going to destroy the farmland and destroy our property values and ruin the rural character of our community!" None of these people were farmers, obviously—they were all rich second homeowners. And I didn't know as much about climate policy then as I do now, but even then I was like, "I'm pretty sure we're gonna need a lot of solar farms. And I'm pretty sure all you're doing is typical NIMBYism."

Curiosity drove him to the town hall meeting advertised in the pamphlet. "After hearing probably ten irate boomers denouncing the solar farm," he says, "I just made the bold move—being the new person in town—to introduce myself and say that I was in favor. And immediately, conversation blew up. And there are all these people accusing me of being a plant from the solar company," Juan-Pablo chuckles. He's familiar with motivated reasoning and political emotions. And at the same time, he tells us, he had a bunch of people—including prominent members of local conservation groups—reach out privately saying, "Hey, thank you for saying that, I'm also privately in favor." "And so afterwards," Juan Pablo says, "I met with a longtime resident who is a retired teacher, who is a beloved figure full of Buddhist serenity, and we started this solar YIMBY group. And pretty quickly, we're able to get a couple of dozen supporters."

These supporters became Friends of Columbia Solar, a grassroots solar advocacy group. The solar farm battle raged in bed and at board across the community. As it became increasingly acrimonious, the YIMBY and the NIMBY groups ultimately decided that collective strategizing was the only way out of the stalemate. Juan-Pablo told us:

> We eventually broke bread with our NIMBY opponents, and worked with a couple of conservation nonprofits and came up with a reworked vision for how the solar farm could still be big, but also be better for the community. And our opponents had had a fairly scorched-earth campaign about "it's too big, it's in the wrong place." And so the fact that we ended up with the same amount of panels in the same spot surprised me. I thought they would have tried to undermine that.

Thoughtful outreach had produced an unlooked-for truce. "And most importantly, I felt that the public would basically be like, 'Oh, it's putting lipstick on a pig," Juan-Pablo said, but "when we presented it, we had almost unanimous positive feedback from the people who've been following the issue in the community. And I think most of the reason was that it was the YIMBYs presenting, then they passed the baton to the NIMBYs, and it was very united." For this small town, the coalition-building among neighbors was crucially important.

"The fact that it really came from the community mattered *way* more than any of the substance," Juan-Pablo concluded. "Which, frankly, surprised me . . . just because I thought, like—people have hard interests on the line. And so I thought that would matter more. But it was really kind of amazing to be like, 'Okay, we were yelling at each other in the press a year ago. Now, we just released something that is actually better, and, in fact, seems to have a consensus.' Is this actually democracy happening?"

Overall though, Juan-Pablo believes, it's time to talk about scale as if we mean it—in solar and everything else. By his reckoning, in order to reduce carbon emissions from 200 million tons annually to net zero by 2050, New York State will need to build 2,000 more wind turbines than they already have, as well as 120 square miles of solar panels. Even if they cover one in five buildings, landfills, and parking lots first, they'll still need forty more miles of solar panels to go on rural land, like that medium-size installation sited for Copake.[9] And every state will need to make similarly divisive zoning decisions.

This means confronting the class interests inherent in interventions like solar farms in wealthy upstate communities. "I think we need to say no to neo-pastoral environmentalism," he says firmly. "When people treat climate solutions aesthetically, they are not taking the problem seriously. And the reality is, we're going to need hundreds and hundreds of very large solar farms and windmills everywhere. People have a visceral reaction against that, and then they twist themselves into pretzels figuring out how to be against it."

We can all recognize by now when a political argument is reverse engineered from a yucky feeling. "The point of climate solutions isn't for you to feel good," he says, a little exasperated, "They're to fix the fucking problem. And so you should always ask: 'what do we need?' And not 'what do I want?'" This brings us back to the old climate taboo problem of "we care, but we don't share." According to ISO New England, an independent non-profit that oversees the region's energy infrastructure, the biggest single barrier to a full clean electricity transition is disputes over where to put transmission lines. We may want renewables, but very few of us are members of grassroots pro-solar, pro-wind, or pro-new transmission line groups. Yet! Realizing that solar advocates are massively outnumbered by solar NIMBYs, even in the left-leaning Hudson Valley, set Juan-Pablo on course to develop WinClimate, which takes lessons learned in local struggles and scales them to state and national policy proposals. And so we proceed to . . .

STATE/REGIONAL

At this point we observe that as an organization's scale gets bigger, there's more likely to be at least a handful of part-time staff to sustain it. It's unusual for projects to get this big without some degree of professional support, especially as they expand to cover larger geographic areas. That doesn't mean that all—or even most—activism turns into wage labor, but volunteer organizers may raise funds or apply for grants in order to hire bookkeepers, experienced campaign managers, paid designers and tech support, legal counsel, and so forth. This is especially true (and potentially tricky) in the case of for-profit projects: when small businesses and co-ops and the like succeed on the local level, the option to scale up presents endless compromises between what is right and what is possible.

But the uptick in compromises and complexity can be very worth it. If we've observed anything in the last few years, it's that states are the hotbed of effective climate action. State campaigns across the country have codified legit climate goals (New York's Climate Leadership and Community Protection Act, and Rhode Island's Act on Climate, for instance); they have overturned antiquated composting bans, and in some places banned organic matter from landfills (essentially mandating compost programs into existence). Other state campaigns aim for 100 percent renewable electricity by 2030 (a measure passed, for instance, in Rhode Island in 2022).

Food Systems

"Food intersects with everything that we do," says Christopher Bradshaw, executive director of food-based nonprofit Dreaming Out Loud. "It impacts every other sector: corporate agribusiness is the biggest polluter in the world. Energy usage in the production of food and transportation of the food is a major factor in our climate. The cows and their methane! Food, it just touches everything. Food is a major aspect of forming our identity. Food's been used as a weapon. Food can be healing, food can be powerful in so many ways."

Food is the first fuel of human life. It represents physical sustenance, culture, habits, values, and love. It is also ensnared in a massive, industrialized economy, and both the economy and food production itself are highly vulnerable to climate change.

Cooperative and localized networks are emerging throughout the country to shift more food to be grown nearer where it's eaten. Those efforts run the gamut: they include scrappy community gardens run by neighbors, medium-sized nonprofit organizations that coordinate distribution, full-scale farmers' markets, and everything in between.

Founded by Christopher Bradshaw in 2008, Dreaming Out Loud aims to rebuild urban, community-based food systems through cooperative social enterprise. They work on increasing access to healthy food, improving community health, supporting entrepreneurs and cooperatives from low-income communities; and creating opportunities for at-risk residents to earn sustainable, family-supporting wages, and build wealth.

Meghan first encountered Dreaming Out Loud when she met Tilman Gerald, the organization's development director, on a tour of a community farm in Washington, DC. The farm that day was a soggy oasis in the heart of the capital, a highly segregated city that is home to many long-term communities as well as legions of temporary inhabitants who come and go with administrations in the form of political staffers. While the farm was a beautiful example of a local Big Yes, Dreaming Out Loud wants to change the way people grow, buy, sell, and eat food in the whole region. We spoke with Tilman and Christopher a few weeks later to learn more about how they understand food justice and how it intersects with racial justice in the context of the broader transition. Christopher explains how food has always been used to exert political power:

> Frederick Douglass talked about the ways in which slave masters would utilize food as a weapon against our enslaved ancestors, whether that was by coercion or denial. Even in modern times, you can look at the ways in which food has been used as a weapon. We don't just produce a lot of grain so that we can feed people and eliminate poverty, that's not the point of it. We produce and subsidize a lot of grain so that we can be at the edge of American soft power or Russian soft power in influencing the developing world as a means of controlling politics and this big picture.

But from this history, he is quick to point out, comes joy as well: "Just as much as food was used as a tool during enslavement, so was the joy of us being able to take the scraps and transform them into soul food and have something to rebuild our culture and convene and have family join in," he says. "We want to be able to rekindle those connections to food. Destigmatize farming and food work, and prioritize workers, and all the things that we've seen this system fail to do—or intentionally do in order to harm or extract."

The challenge, then, is twofold: changing the economy of food, and deepening food cultures in their most physically and socially nourishing forms.

Cracking this nut means thinking differently about food in cities especially. "The interplay between rural and city is one that needs nurturing," Christopher says. "We do think about food production as being a rural thing—and it is, at scale. But so many aspects of our food system necessarily involve cities and urban life, particularly the market. And so most of your eaters—by nature of that dynamic of urban/rural—are going to be in the cities." He nods, then continues: "And so you don't want a situation where the capital is extracted from the rural, and the rural folks are not seeing a return on their labor. And you don't want a situation where folks in the cities are just looked at as the eaters, as the consumers. You want to have a reciprocal ownership relationship [so] we can build out a more circular and robust regional food economy."

One of the ways that this can happen is by a process called procurement. If you've never heard this term you're not alone, but it's a good one to learn now because procurement is a big, underused lever that can change food systems. Large organizations—a municipality, a corporation, or a school, for example—have procurement budgets. This is basically operations money for everything from labor contracts for building staff to bulk toilet paper. But a lot of that budget is for food, whether school lunches or meals for government staff. Any food producer whose products are written into the budget has a reliable purchaser and reliable income, usually for years. When cities direct more of their procurement budgets to local farms and regional producers, then local food systems can occupy a bigger share of the food economy. "When you are able to source from regional farmers," Christopher says, and "rural farmers are receiving that fair price so that they can pay their workers, you can see the mutual benefit and the mutual elevation of those two, country mouse and city mouse.

Working with our country cousins to get good food into the cities and finding new channels for them to generate revenue, we are able to enter into [more] equitable relationships."

Changing procurement can be a people-powered project with big payoffs: Your group could pressure your workplace, university, city, and children's school to source sustainable locally grown ingredients. If there is a Farm-to-School program in your district, it may need support.

To Christopher, the Russian invasion of Ukraine in 2022 illustrated our need to create more resilient local food systems. We know that because of climate change, human conflict and food shocks are on the rise. Russia is the world's largest grain producer, waging war with the fourth- or fifth-largest grain producer. "Any disruption there is going to be a delayed reaction, as the world goes through its reserves of whatever grains," says Christopher. "In the initial disruptions you'll feel increased prices. Even though that's a negative thing, especially for folks that already cannot afford food, it also alerts us that a different solution is needed, that a different system is needed."

Out of crisis, in other words, comes opportunity—*if* we seize it. "So can we take that alert and turn it into action?" Christopher asks. He hopes so: "Regenerative agriculture and regionalized, localized food systems are both restorative and reparative of the climate and of people. But are there also economic opportunities that could more holistically develop our communities?" The video image loses, then regains, clarity as he takes a moment to think. "Those are the things that I hope will happen and we hope that we can be a part of bringing that message to folks in building that power."

We asked Christopher how he might advise someone who cares about climate but is seeking an entry point. You may not be surprised by now to hear another seasoned organizer tell us: find others. He says,

It can seem overwhelming because it *is*. It's overwhelming when you think that you're alone. So you can find places to connect with people that are pushing back, that need solidarity in someone shoulder-to-shoulder with them. Come volunteer. Plug into what we're doing at the farm. It can also be showing up to your civic association or—Here we have ANCs, the Area Neighborhood Commissions, which are little micro-units of DC government—to chime in on issues that are pertinent to your community. It could be joining a group that is organizing a direct action. If you have skill sets, like if you're writing press releases and have media skills and there's a climate action group, well, join that group and help them with a press release! There's so many opportunities, but we can't let the scale of the challenge freeze us and allow us to settle into inaction.

Real Pickles

Real Pickles is a worker-owned cooperative in Western Massachusetts that produces organic, naturally fermented, and raw plant-based food. As a for-profit business, they are using a different strategy to attain allied goals of transforming the food system and economy. As their name suggests they are best known for their pickles, and their folksy mason jars can be seen in shops throughout New England and New York. Dan Rosenberg, bespectacled and with bushy black hair, founded Real Pickles in 2001 at the age of twenty-four. He had emerged from college, which, he said, had been a "pretty transformative experience for me in just understanding the size and nature of society's challenges. I came out of that educational experience pretty focused on organic agriculture and local economics and relocalizing the food system." As a New Englander, he started thinking about what kinds of foods enabled people in his region to eat locally throughout the cold winter . . . and landed on pickles.

As he planned the business, he made several core commitments that reflected his climate and food ethic: that the company would only source vegetables from the Northeast, and that it would only sell its products within the Northeast. "I had," Dan says, "A very strong conviction that the world needs to move away from economic globalization rather than further in that direction, and certainly in the food system." Dan was soon joined in the business by Addie Rose Holland and, he says, "we chose New England, New York, New Jersey, Pennsylvania. That was going to be [our] region. And we were going to stick to it! And part of holding the line was the messaging about it: having a clear way of being able to state it and frame what we were doing." He shrugs and laughs. "And it doesn't really matter if our products go into Maryland, not really. But we needed to make a choice for it to really work, to try to create this model and promote this idea of a regional food system."

They have held the line throughout twenty years of growth and transformation, despite colliding with the norms of the food distribution system along the way. "The scale felt reasonable from a social standpoint and social change standpoint, and it also works functionally," he tells us. But,

> And as we got bigger, we realized we needed to start relying on distributors to get our food where it needed to go, but then distributors have their own particular maps. To this day, we don't have great coverage for getting our products to various parts of Pennsylvania and New Jersey. And one of those reasons is—well there's some great distributors in that region, but they sell all the way to Washington DC and we don't really want to work with them as a result. Certainly, we hit up against customers' expectations about our global capitalist system that we live in, where customers expect they can be able to buy anything, anytime.

The market has what he calls some "crazy externalities" that show up in the price of food. His observations echo some of Christopher Bradshaw's. For instance, even the local grocery co-op (located just down the street from Real Pickles) has "for years, struggled with being able to buy our sauerkraut to use on their deli sandwiches. Because they pay less per pound for their bucket of their *finished* sauerkraut than we do for our raw cabbages from our local organic farms!" He shakes his head and sighs, and laughs, and sighs again.

Ten years into the operation, Dan and Addie decided they wanted to move toward a worker-owned co-op model. They saw this as a way to strengthen and codify their labor values against a punishing, capitalized market, in which actors like the large-scale distributor United Natural Foods were eroding the base. In their view, for the natural organic food businesses over the last twenty to thirty years, "the typical storyline is: business succeeds and does well, ends up selling out to some big corporation. And whatever they might have had for a strong social mission is suddenly up for grabs. Who knows? Most likely going to be weakened, if not disappear altogether." They did not want this to happen to Real Pickles, and so the pair made a decision to enshrine their Yes in the business structure that they were establishing:

> We wanted to *not* go the route of selling out. Instead, workers would be part of the decision making and the ownership, and benefiting from any profits that might accrue. The business is going to stay rooted in the community where people's jobs are. Which is a great upside to worker co-ops—no set of workers are going to vote to ship their jobs out to some other place! . . . There's no 100 percent solid way of preserving a social mission and making sure it doesn't get diluted or abandoned. But a co-op inscribes it in the co-op's governing documents and makes it difficult to change, which seemed to be about the best opportunity out there.

Over time, and even as worker-members have cycled in and out of Real Pickles, the co-op model has enabled them to protect the integrity of their goal and their vision, and become a brick in the building of the local food movement in New England.

NATIONAL

At the national level, many climate organizations are movement-building for political and social action. Here we take a look at groups working to build better cultural infrastructure—specifically, groups working to shift the financial system (one of the core enablers of the climate crisis) and the legal frameworks around parenting.

One thing to note here: even more so than at the state level, national-level organizing can require professionalization and money. It's just harder to push for better rail service or a change in federal policy without those things. But it's important to acknowledge that national organizations or coalitions are often the consequence of groups organizing locally, then linking up regionally, until the institutional shift they're working for is recognized nationally, whether they have offices in every state or not.

Capital Good Fund

The global flow of capital is one of the biggest forces driving the fossil-fueled economy—it makes money unaccountable to communities, corporations unaccountable to people and increasingly, to nations; it locks retirees into terrible investments (of coal, gas, weapons, blood diamonds), and it is designed to be difficult to understand. It affects the urban and the rural. The speed at which financial markets operate, and the demand for quick and big returns on investment, propel neoliberalism's destructive tendencies. As with politics, money and financial markets are entwined in the business of our day-to-day; ignoring that won't make it go away.

Every just transition plan needs an explicit understanding of money.

The Capital Good Fund defines itself as a community development financial institution; it offers small loans to families, helping to create pathways out of poverty and build a green economy through what they call "inclusive" financial services. This is a group that, like the divestment activists of the last chapter, operates squarely within the existing status quo, in order to move that status quo. They are using the tools of finance to change the financial system.

Andy Posner, the founder of Capital Good Fund, describes the personal arc that led him to this work. As a college student during the war in Iraq, he joined protests. "We were chanting "no war for oil!'," he told us. "Whether or not the war was for oil, I don't know. But nevertheless, that got me interested in the issue of, okay, we have a climate crisis, we have geopolitical issues around fossil fuels."

He became a committed cyclist for starts, even biking across the country. But he quickly became aware that cycling alone wouldn't change systems, and started taking a closer look at the intersections of climate, race, and poverty. "This was early 2000s," he tells us during a video interview, "And so at the time, the environmental movement was very white, still very focused on saving trees." During the 2008 financial crisis, some of the pieces started to fall into place for him. "I didn't really understand the connection between Lehman Brothers going under and a poor person getting a loan foreclosed on," he says. He began to explore the financial system, learned about redlining, predatory lending, and all the ways that money locks in inequality. He got curious about how you could use financial services to advance racial justice, economic justice, and climate action.

Apart from predators like payday lenders, Andy saw very few institutions that were providing financial services to struggling families. And so he decided to launch the Capital Good Fund. "The statement is to create pathways out of poverty and advance a green economy through inclusive financial services," he told us. "And that's taken the form of two different types of consumer loan products. Immigration loans and just small-dollar personal [loans] that are an alternative to predatory products."

"The financial rules of the game suck," Andy says, echoing Juan-Pablo's view on the user experience of activism. "The kinds of returns that are affected by the market are only possible through extractive and/or predatory practices, so many of which should not be legal."

For him, the basic premise of the market is geared against health and social justice—this is true for whole countries as well as for individuals. Just as a poor American may have to pay an exorbitant rate because of a low credit score due to circumstances beyond their control—illness or divorce maybe—a poor nation may also see its borrowing costs soar as a result of a climate change-fueled disaster. The cruel irony is that nations paying the highest interest tend to have the lowest emissions. They are experiencing the most severe consequences and are least able to adapt, because they can't access the money. This is the same phenomenon that low-income Americans experience: the average family making $25,000 a year spends as much on financial services (interest, bank fees, etc.) as it does on food, about 10 percent of income.[10] If this kind of predatory lending were illegal, Andy contends that the market would have to reorient its expectations and it would create a more level playing field—even within the context of capitalism itself: "I think you need to regulate [those extractive practices] out of existence."

Andy tells us that the emphasis on immigration loans came out of focus groups held in his community: "People kept mentioning that that was one of their needs. Green card, asylum, citizenship: they didn't have the money." And, he says, immigration reform *increases* the need for capital for legal expenses. "In America, he or she who has the best attorney often wins, and that costs money. And so the wealthy who can afford it are going to get their green card and their family petitions, and low income people often lose and go in the shadows or get deported."

But small-dollar loans in a financial system rigged for the rich and the fossil-fuel economy is still a tough nut to crack. He continues,

> Small dollar loans are impossible to do economically and equitably. And they are usually a Band-Aid. While we continue to offer those, we've pivoted toward a focus on larger loans, loans that are actually not so much a Band-Aid, but more of a solution. In other words, $300 to catch up on rent because you can't afford rent because you don't make enough money to afford housing, and after that loan you still can't afford housing. A $5,000 loan to become a citizen means we still have a broken immigration system. But still that $5,000 is not a Band-Aid, it's a solution to your particular situation.

But the other big move, for the Capital Good Fund—and indeed, the reason we initially asked to talk with them—is because their lending programs are aimed squarely at energy efficiency, solar, and other conversion programs for underserved homeowners. We need to decarbonize all our homes, and quickly, from solar panels to improvements to homes' thermal envelopes to replacing fossil-fueled heating and cooling systems with efficient, electric systems and appliances, like heat pumps, water heaters, inductions stoves. But these are expensive improvements, and the cost still largely falls on individual home residents. The Capital Good Fund is part of an effort to transform weatherization and electrification from a leisure pastime for the wealthy into a universally accessible and incentivized good.

Andy points to some of the existing programs in New England as an example: "There are public programs. For example, in Rhode Island and Massachusetts, there's the Heat Loan program to help people install heat pumps, which is a really cool program. It takes a surcharge in everyone's electric bill, and puts it into a pool of money. And then the pool of money buys down the interest rate to zero, so the homeowner pays no interest on their heat loan."

Within the existing economy, these midrange financing systems are key to more people joining the energy transition, and in this case, into electric heating and cooling. This kind of loan program is nerdy and can often be overlooked as not sufficiently "transformational"—especially because the middle class often has access to financing options already. But this is a game changer for old housing stock, for people with respiratory problems, and for people just barely hanging on to their homes in the first place. We asked Andy about the details of what he's doing and how to communicate it. He agreed that it's difficult:

> There's the big sexy stuff, like Medicare for all, which is cool. But there's also the changing eligibility requirements for Medicare, or closing a coverage gap or making the section 25D solar tax credit refundable. Those things don't make it into the public, popular zeitgeist and probably why Democrats don't get lauded for it. But that nuanced stuff is where it's *actually possible to get wins*, and even bipartisan wins. A lot of the crazy [politicians], they say the crazy stuff for their base. But the base doesn't pay attention to section 25D refundability. They can pass that kind of stuff quietly. And there's a lot of room for progress in those areas.

As Andy is suggesting here, part of the Big Yes is about process—in this case, about financial process. It's about things that move the needle: sometimes moving the needle is flashy and fun, and very often it's not. The grunt work of building an alternative system should not be confused with incrementalism. If we're interested in harm reduction, and in ending extractive and exploitative systems, it's clear that we need thoughtful, equitable alternatives in their place. The point is that even if our Big No is "down with corporate greed!," our strategy can ensure that the end of corporate greed doesn't wallop the people it's already harming on its way out the door.

The Black Maternal Health Momnibus Act (or, the Momnibus)

As we've been saying, many Big Yeses succeed within bigger institutions—or at least, bigger institutions can pull big levers more easily. Here we return to the visionary national policy proposal, known colloquially as the Momnibus, that Dr. Adelle Monteblanco mentioned in chapter 1. First introduced by US Congresswoman Lauren Underwood, Congresswoman Alma Adams, Senator Cory Booker, and members of the Black Maternal Health Caucus in 2021, the Momnibus builds on existing legislation and holistically addresses the maternal health crisis in the United States. The individual bills that make up the package are, themselves, being heard in Congress and have not yet passed as of this writing. They are, however, a stunning glimpse of the world that good policy can envision.

As a refresher: In the United States, the wealthiest nation on earth, moms are dying at the highest rate in the developed world. Each of the twelve titles of the Momnibus was introduced as a stand-alone bill by a member of the Black Maternal Health Caucus in 2021, as a way to approach different facets of the maternal health crisis.[11] It's worth summarizing those bills in some detail, as they exemplify the kind of policy that is both intrinsically good, and a model for policies that states, groups, organizations, and even hospitals could create if they wanted to.

The first bill, the *Social Determinants for Moms Act*, makes investments in things that influence gestational health outcomes, like housing, transportation, and nutrition. This bill provides funding for safe, stable, quality housing for pregnant and postpartum people. It explores the transportation barriers that prevent people from attending pregnancy care appointments and accessing social services; it extends WIC eligibility periods for new moms so that they can access nutritious foods; it creates and funds programs that deliver nutritious food, formula, clean water, and diapers to pregnant and postpartum people in food deserts.

The second bill, the *Kira Johnson Act*, tackles racial bias within healthcare systems. It provides funding to community-based organizations that are working to improve maternal health outcomes for Black pregnant and postpartum people, as well as birthing people from other underserved communities.

The *Protecting Moms Who Served Act* addresses the unique maternal health risks facing veterans, including PTSD and depression. This bill would coordinate pregnancy care at VA facilities, including community resources for housing, nutrition, and employment status. It would identify mental and behavioral health risk factors in the prenatal and postpartum periods, so that veterans get the treatments they need.

The *Perinatal Workforce Act* aims to grow and diversify the perinatal workforce, ensuring that everyone in the United States receives culturally appropriate pregnancy care and support. The bill would fund programs to grow and diversify the maternal health workforce, increasing the number of nurses, physician assistants, and other perinatal health workers who moms can trust throughout their pregnancies, labor and delivery, and the postpartum period.

The *Data to Save Moms Act* responds to the urgent maternal health crisis among Native Americans. This legislation commissions the first-ever comprehensive study to understand the scope of the Native American maternal health crisis, and provides funding to establish the first Tribal maternal mortality review committee.

The *Moms Matter Act* supports moms with maternal mental health conditions and substance use disorders. It invests in community-based programs that provide mental and behavioral health treatments and support, including group prenatal and postpartum care; collaborative maternity care; initiatives to address stigma and raise awareness about warning signs for maternal mental and behavioral health conditions; and suicide prevention programs.

The *Justice for Incarcerated Moms Act* funds exemplary care for pregnant and postpartum people who are incarcerated. The bill also commissions a study to understand the scope of the maternal health crisis among incarcerated people and to make recommendations to prevent death and severe maternal morbidity in American prisons and jails. The bill ties federal funding for prisons and jails to prohibitions on the use of shackling during labor and delivery. This is actually what the bill does, but legalese makes it hard to understand as written: It doesn't *prohibit* shackling, but it won't fund institutions that shackle.

The *Tech to Save Moms Act* invests in the integration and development of telehealth and other digital tools to reduce maternal mortality and severe maternal morbidity, and close racial and ethnic gaps in maternal health outcomes.

The *IMPACT to Save Moms Act* challenges the classist consequences of our healthcare system by promoting equity and quality of care for moms covered by Medicaid. The bill also promotes continuity of health insurance coverage from the start of one's pregnancy through the entire year-long postpartum period and beyond.

The *Maternal Health Pandemic Response Act* makes targeted investments to advance safe and respectful maternity care and improve data collection, monitoring, and research on maternal health outcomes during the COVID-19 pandemic and beyond.

The *Protecting Moms and Babies Against Climate Change Act* takes on climate-related risks, robustly funding initiatives to reduce levels of exposure to extreme heat, air pollution, and other environmental threats to pregnant people, new parents, and their infants. These initiatives include providing people with air-conditioning units, appliances, filtration systems, weatherization support, and direct financial assistance; providing support, including housing and transportation assistance, for people who face the risk of extreme weather events like hurricanes, wildfires, and droughts; promoting community forestry initiatives and tree canopy covers; improving infrastructure and blacktop surfaces; and improving monitoring systems and data sharing for climate change–related risks.

Finally, the *Maternal Vaccination Act* promotes maternal vaccinations to protect the health and safety of moms and babies.

In addition to being a major policy proposal with significant support in Congress, these twelve bills are models for other demands that communities can make, and resources they can create on their own. Contacting your own Congresspeople in support of the Momnibus is perhaps the first move, and a good one! And even if passage of the Momnibus is slow to move through the federal mechanisms, we can work to make the vision that guides these bills a reality in our own states, cities, communities, hospitals, and all the other institutions that comprise our lives. We need not wait for Congress—we can do versions of this elsewhere. Yes!

PLUGGING IN

We hope you find these people and projects as inspiring as we do, but we're aware that it took each activist, legislator, and organizer years of work to realize their Big Yeses. And as we have said repeatedly, we can't go it alone.

Everyone we spoke with began by making a connection to a larger group: whether it was in school, at a farmers' market, by showing up to an event or signing up for a mailing list. Many of them didn't know the journey they were embarking on when they took that first small step. Your first task is *not* to start a national NGO from your kitchen table. Rather, it's to start imagining the world you want to live in. That vision could be as simple as recognizing that the way something is done today drives you crazy.

Then look for a group who is *already working toward it*. Everyone we spoke with could use your help themselves, so if you liked what they're working on, don't be shy about getting in touch.

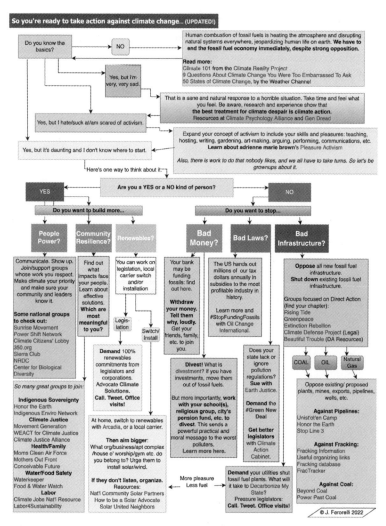

Figure 8.2

Plugging into existing efforts confers other benefits, too: the organizers we spoke with learned their skills from working with other organizers and watching how they do things. And the organizing books and manuals we list are a great resource: many of the titles were recommended by our interviewees.

Be prepared for these campaigns to take months or years—that's the goal of the transformation. Set targets you can meet (e.g., solar panels on the church by the end of the year), and know that achieving one victory earns you the privilege of more organizing.

Finally, we offer a flowchart as a playful way to explore different paths to the conceivable future in figure 8.2.

CONCLUSION

A woman named Kim once told us that in her vision of the conceivable future everyone could "take time to visit a bird or a tree." Like Kourtney's desire to see the stars at night in New England cities, this vision gives us a tantalizing glimpse of where we could be heading. Its sweetness enters us, unexpectedly. It makes us smile. Someone else told us they wanted to see a "maximum wage limit." Still another wanted more park preserves in cities. One woman wanted "a real solution to waste, not just recycling," and another called for universal potable water access.

We hope that as we come to the end of this book we've helped you answer a few important questions: How is the climate crisis shaping *your* life? What is your guiding vision for the future? Who is your family, and how can you enrich those relationships with more mutual understanding and support? Who is your community? How will you join it, tend it, cultivate its change-making power?

If you have been wrestling with whether or not to have a child, we don't presume to have answered that question for you. But we hope that you feel less alone, less burdened with a guilt that doesn't belong to you, and more free to feel. We hope we've created some space, and offered some context and company for you to explore what's important to you, and consider what in your world is worth fighting for.

We want to tell you one more time: climate change is not your fault. It's not your fault for wanting or having kids. It's not your fault for living in a petro-state. Whatever the composition of your family is or will be, whatever loving, imperfect answers you find for yourself to the Impossible Question, it is the fact that we are all asking this question now that describes this climate crisis' moral dimensions. And whatever our families look like, the fact of the question tells us that it is *all* of our responsibility to fight like hell together now. For a future in which no industry's profit is allowed to imperil our collective survival.

In the 1998 movie *Armageddon*, Grace Stamper wonders out loud to her boyfriend, AJ Frost, if anyone else is doing what they are doing right now: in their case, walking animal crackers seductively across each other's bodies in the late-afternoon sunshine. AJ is about to be launched onto an asteroid, and his mission is to drill into it and blow it up, thus diverting its collision-course with Earth. "I hope so," AJ says, looking lovingly at Grace, "otherwise what the hell are we trying to save?"[1]

This clarity cuts through the dated Hollywood schmaltz. In our limited time, only a heartfelt vision—of universal potable water, animal crackers, of generations of thriving families—has the power to support us in this struggle. And this vision is important for more than just making ourselves get up in the morning; we need it to shine a light farther down the path we're walking. Predictions can only be based on precedents, and we are attempting unprecedented things. No one can tell the story of this time until we've written it.

We are lighting a path toward the conceivable future: a country in which we are free and supported to determine the number and spacing of our children, in which we are free and supported to decide to become parents and decide not to; a country that is antiauthoritarian, that practices justice and enshrines human rights. A country that values the health, education, and welfare of all its people equally, above all else. In the conceivable future, these values guide us to decarbonize, transition, and adapt on the atmosphere's time line, rather than accommodating the time line of the current Congress, or some industry's ongoing profits.

Crises can force us to see both who we are, and who we want to become. The very few years left for transformative action give us a unique opportunity to assert the best in human nature and society. We must do that, in part, because we believe it will stabilize the climate enough for life to continue. But we also must do it because—should this be the end of nature's allyship to human survival—it's our responsibility to get right with ourselves and each other. The fantasy of self-sufficiency no longer serves us. We are only sufficient together. There is no point—either moral or strategic—in careening toward a cliff while being cruel.

At its core, climate change is an ethical test, a test of our humanity. We have no time left to be anything other than absolutely loving, absolutely visionary, and absolutely just. We are all parents to the human future, and we are all children, vulnerable in each others' care.

Notes

Introduction

1. The word nibling—one of our favorites—is a gender-neutral term for the child of one's sibling. Niblings is the plural. The counterpoint is pibling, the sibling of a parent.

2. Eliza Relman Hickey Walt, "More than a Third of Millennials Share Rep. Alexandria Ocasio-Cortez's Worry about Having Kids While the Threat of Climate Change Looms," Business Insider, accessed February 12, 2023, https://www.businessinsider.com/millennials-americans-worry-about-kids-children-climate-change-poll-2019-3.

3. Elizabeth Marks et al., "Young People's Voices on Climate Anxiety, Government Betrayal and Moral Injury: A Global Phenomenon," SSRN Scholarly Paper (Rochester, NY, September 7, 2021), https://doi.org/10.2139/ssrn.3918955.

4. Caro Claire Burke, "They Took Away Your Rights Because They Knew Your Husband Wouldn't Care," Substack newsletter, *The Cover Stories* (blog), June 28, 2022, https://caroclaireburke.substack.com/p/they-took-away-your-rights-because.

5. Sanah Ahsan, "I'm a Psychologist—and I Believe We've Been Told Devastating Lies about Mental Health," *The Guardian*, September 6, 2022, sec. Opinion, https://www.theguardian.com/commentisfree/2022/sep/06/psychologist-devastating-lies-mental-health-problems-politics.

6. Erica Chenoweth and Maria J. Stephan, *Why Civil Resistance Works: The Strategic Logic of Nonviolent Conflict* (Columbia University Press, 2011).

7. Rebecca Solnit, "'Hope Is an Embrace of the Unknown': Rebecca Solnit on Living in Dark Times," *The Guardian*, July 15, 2016, sec. Books, https://www.theguardian.com/books/2016/jul/15/rebecca-solnit-hope-in-the-dark-new-essay-embrace-unknown.

8. Rebecca Solnit, *Recollections of My Nonexistence: A Memoir*, Illustrated edition (New York, New York: Viking, 2020).

Part I

1. *The Climate Baby Dilemma*, Documentary (90th Parallel Film and Television Productions, 2022).

2. *The Climate Baby Dilemma*.

3. *The Climate Baby Dilemma*.

4. Stephanie Bailey, "BirthStrike: The People Refusing to Have Kids, Because of 'the Ecological Crisis,'" CNN, June 5, 2019, https://www.cnn.com/2019/06 /05/health/birthstrike-climate-change-scn-intl/index.html.

5. "Our Homeland Is Each Other," NPR, 2018, https://www.npr.org/2018/10 /10/656044800/our-homeland-is-each-other.

6. "Our Homeland Is Each Other."

7. "Our Homeland Is Each Other."

8. mothermade, "Our Voices," 2014, https://www.mothermade.us/2014/01/ our-voices.html.

9. Brady E. Hamilton et al., "Vital Statistics Rapid Release" (Division of Vital Statistics, National Center for Health Statistics, 2016).

10. Dalia Abdel Salam, "Economic Inequality, Violence, and Life in a Changing Climate—10th World Conference of Science Journalists, San Francisco 2017," 2017, http://wcsj2017.org/session/climate-data-conflict/.

11. Marshall Burke, Solomon M. Hsiang, and Edward Miguel, "Climate and Conflict," Working Paper, Working Paper Series (National Bureau of Economic Research, October 2014), https://doi.org/10.3386/w20598.

12. Marshall Burke et al., "Higher Temperatures Increase Suicide Rates in the United States and Mexico," *Nature Climate Change* 8, no. 8 (August 2018): 723–29, https://doi.org/10.1038/s41558-018-0222-x.

13. Bureau of Labor Statistics, "Number of Jobs, Labor Market Experience, Marital Status, and Health: Results from a National Longitudinal Survey" (Washington, DC: Bureau of Labor Statistics, 2021), https://www.bls.gov/news .release/pdf/nlsoy.pdf.

14. Purdue Global, "Generational Differences in the Workplace [Infographic]" (Purdue University, 2023), https://www.purdueglobal.edu/ education-partnerships/generational-workforce-differences-infographic/; Travis Mitchell, "Millennial Life: How Young Adulthood Today Compares with Prior Generations," Pew Research Center's Social & Demographic Trends Project (blog), February 14, 2019, https://www.pewresearch.org/social-trends /2019/02/14/millennial-life-how-young-adulthood-today-compares-with-prior -generations-2/.

15. Angela Fritz, "April Was Earth's 400th Warmer-than-Normal Month in a Row," *Washington Post*, December 3, 2021, https://www.washingtonpost.com /news/capital-weather-gang/wp/2018/05/18/april-was-earths-400th-warmer -than-normal-month-in-a-row/.

16. Gayle Golden, "Climate Change May Cut the Number of Grandparents," Next Avenue, July 12, 2017, https://www.nextavenue.org/climate-change -grandparent/.

CHAPTER 1

1. Tressie McMillan Cottom, "I Was Pregnant and in Crisis. All the Doctors and Nurses Saw Was an Incompetent Black Woman," *Time*, 2019, https://time .com/5494404/tressie-mcmillan-cottom-thick-pregnancy-competent/.

2. March of Dimes, "2022 March of Dimes Report Card" (March of Dimes, 2022), https://www.marchofdimes.org/sites/default/files/2022-11/March-of -Dimes-2022-Full-Report-Card.pdf.

3. Molly Gamble, "34 States Where Child Care Costs More than College Tuition," Becker's Hospital Review, February 17, 2022, https://www .beckershospitalreview.com/rankings-and-ratings/34-states-where-child-care -costs-more-than-college-tuition.html.

4. Molly Redden, "How America's 'Childcare Deserts' Are Driving Women Out of the Workforce," *The Guardian*, January 17, 2018, sec. US news, https: //www.theguardian.com/us-news/2018/jan/16/us-childcare-parenting-mother -load.

5. Kimberly Amadeo, "Living Wage: How Much Do You Need to Live in America?," *The Balance* (blog), 2022, https://www.thebalancemoney.com/living -wage-3305771.

6. Donna St. George, "Teen Girls 'Engulfed' in Violence and Trauma, CDC Finds," *Washington Post*, February 13, 2023, https://www.washingtonpost.com/ education/2023/02/13/teen-girls-violence-trauma-pandemic-cdc/.

7. Tricia Hersey, *Rest Is Resistance: A Manifesto* (Little, Brown, 2022).

8. Olivia McCormack, "Transgender Advocates Say the End of Roe Would Have Dire Consequences," *Washington Post*, May 7, 2022, https://www .washingtonpost.com/politics/2022/05/06/transgender-men-nonbinary-people -abortion-roe/.

9. *The Onion*, "The Pros and Cons of Fracking," *The Onion*, January 29, 2015, https://www.theonion.com/the-pros-and-cons-of-fracking-1819592051.

10. "Fracking Industry Wells Associated with Premature Birth," ScienceDaily, accessed October 26, 2022, https://www.sciencedaily.com/releases/2015/10 /151008110550.htm.

11. Ellen Webb et al., "Neurodevelopmental and Neurological Effects of Chemicals Associated with Unconventional Oil and Natural Gas Operations and Their Potential Effects on Infants and Children," Reviews on Environmental Health 33, no. 1 (March 1, 2018): 3–29, https://doi.org/10.1515/reveh-2017 -0008.

12. Dan Fischer, "Bridgeport Residents Release Balloon Banner at City Hall: 'Fracked Gas Is Environmental Racism,'" Capitalism vs The Climate, 2016, https://ecology.iww.org/node/1492.

13. Stephanie A. Malin and Meghan Elizabeth Kallman, "Enforcing Hopelessness: Complicity, Dependence, and Organizing in Frontline Oil and Gas Communities," *Social Problems*, June 4, 2022, spac032, https://doi.org/10.1093/socpro/spac032.

14. Pauline Mendola et al., "Preterm Birth and Air Pollution: Critical Windows of Exposure for Women with Asthma," *Journal of Allergy and Clinical Immunology* 138, no. 2 (August 2016): 432–440.e5, https://doi.org/10.1016/j.jaci.2015.12.1309.

15. Claire L. Leiser et al., "Acute Effects of Air Pollutants on Spontaneous Pregnancy Loss: A Case-Crossover Study," *Fertility and Sterility* 111, no. 2 (February 1, 2019): 341–47, https://doi.org/10.1016/j.fertnstert.2018.10.028.

16. Chunhua He et al., "Fine Particulate Matter Air Pollution and Under-5 Children Mortality in China: A National Time-Stratified Case-Crossover Study," *Environment International* 159 (January 15, 2022): 107022, https://doi.org/10.1016/j.envint.2021.107022.

17. Tianchao Hu, Brett C. Singer, and Jennifer M. Logue, "Compilation of Published PM2.5 Emission Rates for Cooking, Candles and Incense for Use in Modeling of Exposures in Residences," August 1, 2012, https://doi.org/10.2172/1172959.

18. Dorothy L. Robinson, "Wood Burning Stoves Produce PM2.5 Particles in Amounts Similar to Traffic and Increase Global Warming," *BMJ* 351 (July 14, 2015): h3738, https://doi.org/10.1136/bmj.h3738.

19. Rupa Basu, Brian Malig, and Bart Ostro, "High Ambient Temperature and the Risk of Preterm Delivery," *American Journal of Epidemiology* 172, no. 10 (November 15, 2010): 1108–17, https://doi.org/10.1093/aje/kwq170.

20. Rupa Basu, Varada Sarovar, and Brian J. Malig, "Association Between High Ambient Temperature and Risk of Stillbirth in California," *American Journal of Epidemiology* 183, no. 10 (May 15, 2016): 894–901, https://doi.org/10.1093/aje/kwv295.

21. Emma Bascom, "Extreme Heat May Increase Miscarriage Risk, Study Shows," July 11, 2022, https://www.healio.com/news/primary-care/20220711/extreme-heat-may-increase-miscarriage-risk-study-shows.

22. "What Chicago Can Learn from the 1995 Heat Wave," *What Chicago Can Learn from the 1995 Heat Wave*, July 14, 2020, https://www.npr.org/local/309/2020/07/14/890758229/what-chicago-can-learn-from-the-1995-heat-wave.

23. Aaron S. Bernstein et al., "Warm Season and Emergency Department Visits to U.S. Children's Hospitals," *Environmental Health Perspectives* 130, no. 1 (2022): 017001, https://doi.org/10.1289/EHP8083.

24. Samantha Montano, *Disasterology: Dispatches from the Frontlines of the Climate Crisis* (Toronto: Park Row Books, 2021).

25. United Nations High Commissioner for UNHCR, "Frequently Asked Questions on Climate Change and Disaster Displacement," UNHCR (blog),

2016, https://www.unhcr.org/news/latest/2016/11/581f52dc4/frequently-asked-questions-climate-change-disaster-displacement.html.

26. Roxane Richter and Thomas Flowers, "Gendered Dimensions of Disaster Care: Critical Distinctions in Female Psychosocial Needs, Triage, Pain Assessment, and Care," *American Journal of Disaster Medicine* 3, no. 1 (January 1, 2008): 31–37, https://doi.org/10.5055/ajdm.2008.0004.

27. C.-Y. Cynthia Lin, "Instability, Investment, Disasters, and Demography: Natural Disasters and Fertility in Italy (1820–1962) and Japan (1671–1965)," *Population and Environment* 31, no. 4 (March 1, 2010): 255–81, https://doi.org/10.1007/s11111-010-0103-3.

28. Eric Neumayer and Thomas Plümper, "The Gendered Nature of Natural Disasters: The Impact of Catastrophic Events on the Gender Gap in Life Expectancy, 1981–2002," *Annals of the Association of American Geographers* 97, no. 3 (September 1, 2007): 551–66, https://doi.org/10.1111/j.1467-8306.2007.00563.x.

29. Ophra Leyser-Whalen, Mahbubur Rahman, and Abbey B. Berenson, "Natural and Social Disasters: Racial Inequality in Access to Contraceptives After Hurricane Ike," *Journal of Women's Health* 20, no. 12 (December 2011): 1861–66, https://doi.org/10.1089/jwh.2010.2613.

30. Nahid Rezwana and Rachel Pain, "Gender-Based Violence before, during, and after Cyclones: Slow Violence and Layered Disasters," *Disasters* 45, no. 4 (2021): 741–61, https://doi.org/10.1111/disa.12441.

31. Emily W. Harville et al., "Postpartum Mental Health after Hurricane Katrina: A Cohort Study," *BMC Pregnancy and Childbirth* 9, no. 1 (June 8, 2009): 21, https://doi.org/10.1186/1471-2393-9-21.

32. David L Paterson, Hugh Wright, and Patrick N A Harris, "Health Risks of Flood Disasters," *Clinical Infectious Diseases* 67, no. 9 (October 15, 2018): 1450–54, https://doi.org/10.1093/cid/ciy227.

33. Andrew Freedman, "Zika Cases Prompt CDC to Issue a 1-Mile No-Go Zone for Pregnant Women in Miami," Mashable, August 1, 2016, https://mashable.com/article/zika-outbreak-miami-cdc-travel-advisory.

34. Hagai Levine et al., "Temporal Trends in Sperm Count: A Systematic Review and Meta-Regression Analysis," *Human Reproduction Update* 23, no. 6 (November 1, 2017): 646–59, https://doi.org/10.1093/humupd/dmx022.

35. Levine et al.

36. Levine et al., "Temporal Trends in Sperm Count."

37. Elizabeth Arias and Farida B. Ahmad, "Provisional Life Expectancy Estimates for 2021," Vital Statistics Rapid Release, August 2022, 16.

38. Arias and Ahmad.

39. Mark Olfson et al., "Declining Health-Related Quality of Life in the U.S.," *American Journal of Preventive Medicine* 54, no. 3 (March 1, 2018): 325–33, https://doi.org/10.1016/j.amepre.2017.11.012.

40. Samantha Montano, *Disasterology: Dispatches from the Frontlines of the Climate Crisis* (Toronto: Park Row Books, 2021).

CHAPTER 2

1. Jade S. Sasser, *On Infertile Ground: Population Control and Women's Rights in the Era of Climate Change* (New York: NYU Press, 2018).

2. José A. Tapia Granados, Edward L. Ionides, and Óscar Carpintero, "Climate Change and the World Economy: Short-Run Determinants of Atmospheric CO_2," *Environmental Science & Policy* 21 (August 1, 2012): 50–62, https://doi.org/10.1016/j.envsci.2012.03.008.

3. Sasser, On Infertile Ground.

4. See *Reproductive Rights and Wrongs* by Betsy Hartmann; Haymarket Books, 2013.

5. Betsy Hartmann, "Population Control I: Birth of an Ideology," International Journal of Health Services 27, no. 3 (1997): 523–40.

6. Leyland Cecco, "Human Rights Groups Call on Canada to End Coerced Sterilization of Indigenous Women," *The Guardian*, November 18, 2018, sec. World news, https://www.theguardian.com/world/2018/nov/18/canada-indigenous-women-coerced-sterlilization-class-action-lawsuit; Alexandra Minna Stern, "STERILIZED in the Name of Public Health," *American Journal of Public Health* 95, no. 7 (July 2005): 1128–38, https://doi.org/10.2105/AJPH.2004.041608.

7. Derek Hawkins, "Judge to Inmates: Get Sterilized and I'll Shave off Jail Time," *Washington Post*, October 25, 2021, https://www.washingtonpost.com/news/morning-mix/wp/2017/07/21/judge-to-inmates-get-sterilized-and-ill-shave-off-jail-time/; Christine Dehlendorf and Kelsey Holt, "Opinion | The Dangerous Rise of the IUD as Poverty Cure," *New York Times*, January 2, 2019, sec. Opinion, https://www.nytimes.com/2019/01/02/opinion/iud-implants-contraception-poverty.html.

8. "Abortion and Women of Color: The Bigger Picture," Guttmacher Institute, August 13, 2008, https://doi.org/10.1363/2008.13307.

9. While there are many examples of population programs sterilizing men as well as women, with or without coercion, population programs are overwhelmingly focused on female anatomy and gender-focused on motherhood, women, and girls.

10. Adams Elise, "Voluntary Sterilization of Inmates for Reduced Prison Sentences," *Duke Journal of Gender Law and Policy* 26 (December 20, 2018), https://djglp.law.duke.edu/article/voluntary-sterilization-of-inmates-for-reduced-prison-sentences-adams-vol26-iss1/.

11. "Sierra Club Report Sierra Club Population Report, Spring 1989 - SUSPS - Sierra Club Population Immigration Ballot Question," accessed December 28, 2022, https://www.susps.org/history/popreport1989.html.

12. Martin King Whyte, Wang Feng, and Yong Cai, "Challenging Myths About China's One-Child Policy," *The China Journal* 74 (July 2015): 144–59, https://doi.org/10.1086/681664.

13. "SUSPS—Support US Population Stabilization—Support a Comprehensive Sierra Club Population Policy," accessed December 28, 2022, https://www.susps.org/.

14. Miguel Bustillo and Kenneth R. Weiss, "Election Becomes a Fight Over Sierra Club's Future," *Los Angeles Times*, January 18, 2004, https://www.latimes.com/archives/la-xpm-2004-jan-18-me-sierra18-story.html.

15. Truven Health Analytics, "The Cost of Having a Baby in the United States" (Truven Health Analytics, 2013), https://www.nationalpartnership.org/our-work/resources/health-care/maternity/cost-of-having-a-baby-executive-summary.pdf.

16. Redden, "How America's 'Childcare Deserts' Are Driving Women out of the Workforce."

17. Bill Chappell, "U.S. Births Fell To A 32-Year Low In 2018; CDC Says Birthrate Is In Record Slump," NPR, May 15, 2019, sec. National, https://www.npr.org/2019/05/15/723518379/u-s-births-fell-to-a-32-year-low-in-2018-cdc-says-birthrate-is-at-record-level; Anna Louie Sussman, "Conceiving the Future," accessed December 29, 2022, https://www.nybooks.com/articles/2021/09/23/climate-change-birth-conceiving-future/.

18. Arnstein Aassve et al., "Early Assessment of the Relationship between the COVID-19 Pandemic and Births in High-Income Countries," *Proceedings of the National Academy of Sciences* 118, no. 36 (September 7, 2021): e2105709118, https://doi.org/10.1073/pnas.2105709118.

19. Dianne Lalonde, "Sexist Barriers Block Women's Choice to Be Sterilized," The Conversation, 2018, http://theconversation.com/sexist-barriers-block-womens-choice-to-be-sterilized-99754.

20. Paul Ehrlich, "Paul Ehrlich: 'Collapse of Civilisation Is a Near Certainty within Decades,'" *The Guardian*, March 22, 2018, sec. Cities, https://www.theguardian.com/cities/2018/mar/22/collapse-civilisation-near-certain-decades-population-bomb-paul-ehrlich.

21. Center for Biological Diversity, "Food Waste Is Trashing the Planet," *Biological Diversity* (blog), 2021, http://www.takeextinctionoffyourplate.com/waste/index.html.

22. Jared Starr et al., "Assessing U.S. Consumers' Carbon Footprints Reveals Outsized Impact of the Top 1%," *Ecological Economics* 205 (March 1, 2023): 107698, https://doi.org/10.1016/j.ecolecon.2022.107698.

23. Beatriz Barros and Richard Wilk, "The Outsized Carbon Footprints of the Super-Rich," *Sustainability: Science, Practice and Policy* 17, no. 1 (January 1, 2021): 316–22, https://doi.org/10.1080/15487733.2021.1949847.

24. Aaron R. Brough et al., "Is Eco-Friendly Unmanly? The Green-Feminine Stereotype and Its Effect on Sustainable Consumption," *Journal of Consumer*

Research 43, no. 4 (December 1, 2016): 567–82, https://doi.org/10.1093/jcr/ucw044.

25. Jonathan Cohn, "What The Weird Conservative Fixation On Electric Vehicles Says About Their Politics," Huffington Post, October 7, 2022, https://www.huffpost.com/entry/marjorie-taylor-greene-ev-pete-buttigieg_n_63408787e4b028164534ca74.

26. Katarina Hamberg, "Gender Bias in Medicine," *Women's Health* 4, no. 3 (May 1, 2008): 237–43, https://doi.org/10.2217/17455057.4.3.237.

27. Gabrielle Jackson, "The Female Problem: How Male Bias in Medical Trials Ruined Women's Health," *The Guardian*, November 13, 2019, sec. Life and style, https://www.theguardian.com/lifeandstyle/2019/nov/13/the-female-problem-male-bias-in-medical-trials.

28. Jackson.

29. Rob Haskell, "Serena Williams on Motherhood, Marriage, and Making Her Comeback," *Vogue*, January 10, 2018, https://www.vogue.com/article/serena-williams-vogue-cover-interview-february-2018.

30. McMillan Cottom, "I Was Pregnant and in Crisis. All the Doctors and Nurses Saw Was an Incompetent Black Woman."

31. Seth Wynes and Kimberly A. Nicholas, "The Climate Mitigation Gap: Education and Government Recommendations Miss the Most Effective Individual Actions," *Environmental Research Letters* 12, no. 7 (July 2017): 074024, https://doi.org/10.1088/1748-9326/aa7541.

32. Paul A. Murtaugh and Michael G. Schlax, "Reproduction and the Carbon Legacies of Individuals," *Global Environmental Change* 19, no. 1 (February 1, 2009): 14–20, https://doi.org/10.1016/j.gloenvcha.2008.10.007.

33. Ross, Loretta, Elena Gutierrez, Marlene Gerber, and Jael Silliman. 2016. *Undivided Rights: Women of Color Organizing for Reproductive Justice.* Haymarket Books.

34. Sasser, *On Infertile Ground.*

35. Bond, Toni M. 1994. Quote from press release.

36. Centers for Disease Control and Prevention, "Infant Mortality | Maternal and Infant Health | Reproductive Health," September 8, 2022, https://www.cdc.gov/reproductivehealth/maternalinfanthealth/infantmortality.htm.

37. Alexis Robles-Fradet, "Medicaid Coverage for Doula Care: 2021 State Roundup," *National Health Law Program* (blog), December 8, 2021, https://healthlaw.org/medicaid-coverage-for-doula-care-2021-state-roundup/.

38. Economic Policy Institute, "Child Care Costs in the United States," *Economic Policy Institute* (blog), 2021, https://www.epi.org/child-care-costs-in-the-united-states/.

39. Rasheed Malik and Katie Hamm, "Mapping America's Child Care Deserts," *Center for American Progress* (blog), 2021, https://www.americanprogress.org/article/mapping-americas-child-care-deserts/.

40. Team Warren, "My Plan for Universal Child Care," *Medium* (blog), 2019, https://medium.com/@teamwarren/my-plan-for-universal-child-care -762535e6c20a.

41. Barbara Broman, "National Center on Family Homelessness," American Institutes for Research, 2021, https://www.air.org/centers/national-center -family-homelessness.

42. http://socialsciences.people.hawaii.edu/publications_lib/domestic %20violence%20and%20housing.pdf.

CHAPTER 3

1. Shruti Jolly et al., "Gender Differences in Time Spent on Parenting and Domestic Responsibilities by High-Achieving Young Physician-Researchers," *Annals of Internal Medicine* 160, no. 5 (March 4, 2014): 344–53, https://doi .org/10.7326/M13-0974; US Bureau of Labor Statistics, "American Time Use Survey," 2021, https://www.bls.gov/tus/home.htm.

2. The Impact of the 2015–16 El-nino Induced Drought on Women, *Dynamic Research Journals* (DRJ), *Journal of Economics and Finance* (DRJ-JEF), 2: no. 9 (September 2017) pp: 10–17.

3. AAUW, "The Simple Truth About the Gender Pay Gap: AAUW Report" (American Association of University Women, 2021), https://www.aauw.org/ resources/research/simple-truth/.

4. Jessica Gaitán Johannesson, *The Nerves and Their Endings: Essays on Crisis and Response* (Scribe US, 2023).

5. Heather McMullen and Katharine Dow, "'Too Afraid to Have Kids' —How BirthStrike for Climate Lost Control of Its Political Message," The Conversation, accessed December 6, 2022, http://theconversation.com/too -afraid-to-have-kids-how-birthstrike-for-climate-lost-control-of-its-political -message-181198.

6. Betty Friedan, *The Feminine Mystique* (New York: Norton, 2001).

7. Polly Poskin, "A Brief History of the Anti-Rape Movement" (Illinois Coalition Against Sexual Assault, 2006), http://resourcesharingproject.org/wp -content/uploads/2021/11/History_of_the_Movement.pdf.

8. Susan Brownmiller, *Against Our Will: Men, Women, and Rape*, English Language edition (Bantam Books, 1976).

9. Shelford Bidwell, *Modern Warfare: A Study of Men, Weapons and Theories* (Allen Lane, 1973).

10. Gina Marie Weaver, *Ideologies of Forgetting: Rape in the Vietnam War* (Albany: State University of New York Press, 2010).

11. Cassandra Clifford, "Rape as a Weapon of War and It's Long-Term Effects on Victims and Society" (7th Global Conference Violence and the Contexts of Hostility, Budapest, Hungary, 2008), 13, vaw_rapeasaweaponofwar_stopmodernslavery_may2008_0.

12. Lisa Vetten and Sadiyya Haffejee, "Gang Rape: A Study in Inner-City Johannesburg," *South African Crime Quarterly*, no. 12 (2005), https://doi.org/10.17159/2413-3108/2005/i12a1017; Rachel A. Feinstein, *When Rape Was Legal: The Untold History of Sexual Violence During Slavery* (New York: Routledge, 2018), https://doi.org/10.4324/9781315210285.

13. Roy L. Austin and Young S. Kim, "A Cross-National Examination of the Relationship between Gender Equality and Official Rape Rates," *International Journal of Offender Therapy and Comparative Criminology* 44, no. 2 (April 1, 2000): 204–21, https://doi.org/10.1177/0306624X00442006; James S. E. Herman et al., "Report of the 2015 US Transgender Survey" (Washington, DC: National Center for Transgender Equality, 2016), https://transequality.org/sites/default/files/docs/usts/USTS-Executive-Summary-Dec17.pdf.

14. It's worth considering here too whether some of the harshness of the reception of baby strike/pledge activists was based on the fact that they explicitly *didn't* conform to the ideal of motherhood. This nonconformity was, unsurprisingly, met online with threats of sexual violence.

15. Marcela Perelman and Veronica Torras, "The Mothers of Plaza de Mayo: 'We Were Born on the March,'" *OpenDemocracy* (blog), 2017, https://www.opendemocracy.net/en/mothers-plaza-de-mayo/.

16. Laura J. Shepherd, ed., *Gender Matters in Global Politics: A Feminist Introduction to International Relations*, 2nd edition (London: Routledge, 2014).

17. Mabel Bellucci, "Childless Motherhood: Interview with Nora Cortiñas, a Mother of the Plaza de Mayo, Argentina," *Reproductive Health Matters* 7, no. 13 (January 1, 1999): 83–88, https://doi.org/10.1016/S0968-8080(99)90116-7.

18. Matthew, "History of Mother's Day as a Day of Peace: Julia Ward Howe," The Peace Alliance, May 8, 2015, https://peacealliance.org/history-of-mothers-day-as-a-day-of-peace-julia-ward-howe/.

PART 2

1. "Yale Climate Opinion Maps 2021," Yale Program on Climate Change Communication (blog), accessed October 25, 2022, https://climatecommunication.yale.edu/visualizations-data/ycom-us/.

2. Chenoweth and Stephan, *Why Civil Resistance Works*.

CHAPTER 4

1. William James, *The Varieties of Religious Experience:, The Works of William James* (Cambridge, MA: Harvard University Press, 1985). Chapter 9.

2. Geoffrey Supran and Naomi Oreskes, "Rhetoric and Frame Analysis of ExxonMobil's Climate Change Communications," *One Earth* 4, no. 5 (May 21, 2021): 696–719, https://doi.org/10.1016/j.oneear.2021.04.014.

3. Matthew Schneider-Mayerson, "The Influence of Climate Fiction: An Empirical Survey of Readers," *Environmental Humanities* 10, no. 2 (November 1, 2018): 473–500, https://doi.org/10.1215/22011919-7156848.

4. Paul B. Sturtevant, "Were the Middle Ages Really 'The Good Place'?," *The Public Medievalist*, January 10, 2020, https://www.publicmedievalist.com/medieval-the-good-place/.

5. Sturtevant.

6. "Do You Blast the A/C? Throw out Half Your Lunch? Tell Us Where You Fall Short in Preventing Climate Change," NBC News, accessed March 14, 2023, https://www.nbcnews.com/news/specials/climate-confessions-share-solutions-climate-change-n1054791.

7. Andrea Schneibel, "Relationship Between Low Income and Obesity Is Relatively New," *News*, December 11, 2018, https://news.utk.edu/2018/12/11/relationship-between-low-income-and-obesity-is-relatively-new/.

8. Laura says every COVID death is a policy failure [@TutusNTinyHats], "This Is Especially Tricky for Fat People, Because Most of the 'Sustainable,' 'Eco-Friendly,' Etc Brands Don't Make plus Sizes.," Tweet, Twitter, November 24, 2019, https://twitter.com/TutusNTinyHats/status/1198630825312604161.

9. Bill McKibben, "Stop Talking Right Now about the Threat of Climate Change. It's Here; It's Happening," *The Guardian*, September 11, 2017, sec. Opinion, https://www.theguardian.com/commentisfree/2017/sep/11/threat-climate-change-hurricane-harvey-irma-droughts.

10. Julie Guthman, *Weighing in: Obesity, Food Justice, and the Limits of Capitalism*, California Studies in Food and Culture 32 (Berkeley: University of California Press, 2011).

11. Emily Nagoski and Amelia Nagoski, *Burnout: The Secret to Unlocking the Stress Cycle*, Reprint edition (Random House Publishing Group, 2020).

12. Check out the appendix for some resources we like.

13. Margolin, Jamie. (2019, Jan 19). MY biggest pet peeve: When people tell me it's so cool how "passionate" I am about #climatechange and recommend books and movies [Video of youth activist panel]. Facebook. https://fb.watch/j3vL9sCFy6/

14. Danielle Carr, "Mental Health Is Political," *New York Times*, September 20, 2022, sec. Opinion, https://www.nytimes.com/2022/09/20/opinion/us-mental-health-politics.html.

CHAPTER 5

1. Hilary Rose and Shannon Hebblethwaite, "Expanding the Definition of Family to Reflect Our Realities," The Conversation (blog), 2020, http://theconversation.com/expanding-the-definition-of-family-to-reflect-our-realities-131743.

2. Pew Research Center, "The American Family Today," *Pew Research Center's Social & Demographic Trends Project* (blog), December 17, 2015, https://www.pewresearch.org/social-trends/2015/12/17/1-the-american-family-today/.

3. Hui Liu et al., "Marital Status and Dementia: Evidence from the Health and Retirement Study," *The Journals of Gerontology*: Series B 75, no. 8 (October 1, 2020): 1783–95, https://doi.org/10.1093/geronb/gbz087.

4. Esther O. Lamidi, "Trends in Self-Rated Health by Union Status and Education, 2000–2018," *SSM—Population Health* 11 (August 1, 2020): 100616, https://doi.org/10.1016/j.ssmph.2020.100616.

5. Mieke Beth Thomeer, "Relationship Status-Based Health Disparities during the COVID-19 Pandemic," *Social Currents*, May 13, 2022, 23294965221099184, https://doi.org/10.1177/23294965221099185.

6. Robert J. Waldinger and Marc S. Schulz, "What's Love Got to Do With It?: Social Functioning, Perceived Health, and Daily Happiness in Married Octogenarians," *Psychology and Aging* 25, no. 2 (June 2010): 422–31, https://doi.org/10.1037/a0019087.

7. Mieke Beth Thomeer, Jenjira Yahirun, and Alejandra Colón-López, "How Families Matter for Health Inequality during the COVID-19 Pandemic," *Journal of Family Theory & Review* 12, no. 4 (December 2020): 448–63, https://doi.org/10.1111/jftr.12398.

8. David Brooks, "The Nuclear Family Was a Mistake," *The Atlantic*, February 10, 2020, https://www.theatlantic.com/magazine/archive/2020/03/the-nuclear-family-was-a-mistake/605536/.

9. Richard Fry, "Young Adults in U.S. Are Much More Likely than 50 Years Ago to Be Living in a Multigenerational Household," *Pew Research Center* (blog), 2022, https://www.pewresearch.org/fact-tank/2022/07/20/young-adults-in-u-s-are-much-more-likely-than-50-years-ago-to-be-living-in-a-multigenerational-household/.

10. D'Vera Cohn et al., "Financial Issues Top the List of Reasons U.S. Adults Live in Multigenerational Homes," *Pew Research Center's Social & Demographic Trends Project* (blog), March 24, 2022, https://www.pewresearch.org/social-trends/2022/03/24/financial-issues-top-the-list-of-reasons-u-s-adults-live-in-multigenerational-homes/.

11. Zhen Cong and Merril Silverstein, "Intergenerational Support and Depression Among Elders in Rural China: Do Daughters-In-Law Matter?," *Journal of Marriage and Family* 70, no. 3 (2008): 599–612, https://doi.org/10.1111/j.1741-3737.2008.00508.x; Haena Lee et al., "Multigenerational Households During Childhood and Trajectories of Cognitive Functioning Among U.S. Older Adults," *The Journals of Gerontology*: Series B 76, no. 6 (July 1, 2021): 1161–72, https://doi.org/10.1093/geronb/gbaa165.

12. Brooks, "The Nuclear Family Was a Mistake."

13. David Harvey, *A Brief History of Neoliberalism* (Oxford University Press, 2007).

14. Ellen Barry, "A Massachusetts City Decides to Recognize Polyamorous Relationships," *New York Times*, July 2, 2020, sec. U.S., https://www.nytimes.com/2020/07/01/us/somerville-polyamorous-domestic-partnership.html.

15. Imani Perry, "Why Is 'Auntie' a Controversial Word?," Unsettled Territory/ *The Atlantic*, April 6, 2022, https://newsletters.theatlantic.com/unsettled -territory/624dc597c42c790021169148/auntie-word-ageism-black-women/.

16. The Stepfamily Foundation, "Stepfamily Statistics," The Stepfamily Foundation Inc., 2022, https://www.stepfamily.org/stepfamily-statistics.html.

17. R. J. Reinhart, "Global Warming Age Gap: Younger Americans Most Worried," Gallup.com, May 11, 2018, sec. Politics, https://news.gallup.com/poll /234314/global-warming-age-gap-younger-americans-worried.aspx.

18. Emily Brandon, "Why Older Citizens Are More Likely to Vote," *US News & World Report*, 2020, https://money.usnews.com/money/retirement/aging/ articles/why-older-citizens-are-more-likely-to-vote.

19. Susie E. L. Burke, Ann V. Sanson, and Judith Van Hoorn, "The Psychological Effects of Climate Change on Children," *Current Psychiatry Reports* 20, no. 5 (April 11, 2018): 35, https://doi.org/10.1007/s11920-018-0896 -9; Gina Martin et al., "Review: The Impact of Climate Change Awareness on Children's Mental Well-Being and Negative Emotions—A Scoping Review," *Child and Adolescent Mental Health* 27, no. 1 (February 2022): 59–72, https://doi .org/10.1111/camh.12525.

20. Sanchi Ravishanker, "Children and Climate Change," 2020, https://gdc .unicef.org/resource/children-and-climate-change.

21. Golden, "Climate Change May Cut the Number of Grandparents."

CHAPTER 6

1. See Robert Putnam, *Bowling Alone*.

2. Doug McAdam, *Political Process and the Development of Black Insurgency, 1930–1970* (University of Chicago Press, 1982); Aldon D. Morris, *The Origins of the Civil Rights Movement: Black Communities Organizing for Change* (Simon & Schuster, 1986).

3. Cindy Kallet et al., Farthest Field, Heartwalk (Overall Music OM-3, 1993).

4. Doug McAdam, "Recruitment to High-Risk Activism: The Case of Freedom Summer," *American Journal of Sociology* 92(1)(1986): 64–90.

5. Florence Passy and Marco Giugni, "Social Networks and Individual Perceptions: Explaining Differential Participation in Social Movements," *Sociological Forum* 16(1)(2001):123–53.

6. Roger Gould, Multiple Networks and Mobilization in the Paris Commune, 1871. *Am. Sociol. Rev.* 56, 716729 (1991).

7. Richard Flory and Donald E. Miller, *The Embodied Spirituality of the Post-Boomer Generations: A Sociology of Spirituality* (Routledge, 2016), 217, https://doi .org/10.4324/9781315565231-12.

8. Casper Kuile and Angie Thurston, "How We Gather" (Cambridge, MA: Harvard Divinity School, 2016), https://caspertk.files.wordpress.com/2015/04/how-we-gather.pdf.

9. kyodergrist, "Greentrolling: A 'Maniacal Plan' to Bring down Big Oil," Grist, November 19, 2020, https://grist.org/energy/greentrolling-a-maniacal-plan-to-bring-down-big-oil/.

10. Sammy Nickalls, "Here's Why All Your Friends Are Deleting Their Uber Accounts," *Esquire*, January 29, 2017, https://www.esquire.com/news-politics/news/a52652/heres-why-uber/.

11. Priya Parker's *The Art of Gathering* offers some great models for thinking through convening.

12. Casper ter Kuile, *The Power of Ritual: Turning Everyday Activities into Soulful Practices* (New York: HarperOne, 2020).

CHAPTER 7

1. Amy Westervelt [@amywestervelt], "Periodic Reminder That the Window of Opportunity for Moderate Solutions on Climate Was Slammed Shut in Favor of Oil Company Profits for 30 Years and Now All of the Options Are Extreme.," Tweet, Twitter, September 17, 2021, https://twitter.com/amywestervelt/status/1439001803878666240.

2. Sarah Hurtes, "Women Who Risk Everything to Defend the Environment," *Harper's Bazaar*, November 30, 2018, https://www.harpersbazaar.com/culture/features/a22737480/leeanne-walters-cherri-foytlin-tara-houska-women-environmental-activists-face-danger/.

3. Nadia Y. Bashir et al., "The Ironic Impact of Activists: Negative Stereotypes Reduce Social Change Influence," *European Journal of Social Psychology* 43, no. 7 (2013): 614–26, https://doi.org/10.1002/ejsp.1983.

4. Gene Sharp, "198 Methods of Nonviolent Action" (The Albert Einstein Institute, 1973), https://www.jeffreythompson.org/downloads/198MethodsOfNonviolentAction.pdf.

5. The Success of Nonviolent Civil Resistance: Erica Chenoweth at TEDxBoulder, 2013, https://www.youtube.com/watch?v=YJSehRlU34w.

6. "Guidebook," Failing to Save the Earth, accessed January 1, 2023, https://www.failingtosavetheearth.com/books.

7. Ben Collins, "Meet the Man Who Beat the KKK With a Tuba," The Daily Beast, July 22, 2015, sec. us-news, https://www.thedailybeast.com/articles/2015/07/22/meet-the-man-who-beat-the-kkk-with-a-tuba.

8. Jedediah Purdy, "Occupy Your Voice," *HuffPost*, February 2, 2012, sec. Politics, https://www.huffpost.com/entry/occupy-your-voice_b_1249398.

9. Margaret Klein Salamon and Molly Gage, *Facing the Climate Emergency: How to Transform Yourself with Climate Truth* (New Society Publishers, 2020).

10. Kate Schapira, "About," *Climateanxietycounseling* (blog), April 28, 2014, https://climateanxietycounseling.wordpress.com/about/.

Notes

11. Ruth McDermott-Levy, Nina Kaktins, and Barbara Sattler, "Fracking, the Environment, and Health," *AJN: The American Journal of Nursing* 113, no. 6 (June 2013): 45, https://doi.org/10.1097/01.NAJ.0000431272.83277.f4.

12. Elaine L. Hill and Lala Ma, "Drinking Water, Fracking, and Infant Health," *Journal of Health Economics* 82 (March 1, 2022): 102595, https://doi.org/10.1016/j.jhealeco.2022.102595.

13. "Delaware River Frack Ban Coalition 'Deeply Disturbed' by DRBC Failure to Fully Ban Fracking Activity," Delaware First Media (Delaware Public Media, December 29, 2022), https://www.delawarepublic.org/delaware-headlines/2022-12-29/delaware-river-frack-ban-coalition-deeply-disturbed-by-drbc-failure-to-fully-ban-fracking-activity.

14. "Push for Fracking Ban in Delaware River Watershed Continues," 47abc, December 27, 2022, https://www.wmdt.com/2022/12/push-for-fracking-ban-in-delaware-river-watershed-continues/.

15. Bill McKibben, "At Last, Divestment Is Hitting the Fossil Fuel Industry Where It Hurts," *The Guardian*, December 16, 2018, sec. Opinion, https://www.theguardian.com/commentisfree/2018/dec/16/divestment-fossil-fuel-industry-trillions-dollars-investments-carbon.

16. "Impacts of the Fossil Fuel Divestment Movemen—Noam Bergman," Centre on Innovation and Energy Demand (CIED), accessed February 28, 2023, http://www.cied.ac.uk/publication/impacts-divestment-movement/.

17. Bill McKibben, "Global Warming's Terrifying New Math," *Rolling Stone* (blog), July 19, 2012, https://www.rollingstone.com/politics/politics-news/global-warmings-terrifying-new-math-188550/.

18. Kathy Hipple, "IEEFA Update: Fiduciary Duty and Fossil Fuel Divestment," October 2, 2019, https://ieefa.org/resources/ieefa-update-fiduciary-duty-and-fossil-fuel-divestment.

19. Elizabeth Shogren, "College Divestment Campaigns Creating Passionate Environmentalists," NPR, May 10, 2013, sec. Environment, https://www.npr.org/2013/05/10/182599588/college-divestment-campaigns-creating-passionate-environmentalists.

20. "Pluralistic Ignorance: Definition & Examples," accessed November 5, 2022, https://www.simplypsychology.org/pluralistic-ignorance.html.

21. Rev. Dr. Martin Luther King, "Letter from a Birmingham Jail," April 16, 1963, https://www.africa.upenn.edu/Articles_Gen/Letter_Birmingham.html.

22. steve albini [@electricalWSOP], "People Are Dunking on These Absurdist Demonstrations but I Think They're Kinda Cool. The Point They're Making Is That Your Most Priceless Shit Is Worthless on a Dead Planet, and While They're Not Going to Change Any Minds It's a Way to Get That Idea into the Discourse.," Tweet, Twitter, October 14, 2022, https://twitter.com/electricalWSOP/status/1580950091940034560.

CHAPTER 8

1. Syd [@SydneyAzari], "I Want Future Children to Grow up in a World Knowing That the People Alive Today Loved Them so Much They Were Willing to Risk Everything so That They Could Live.," Tweet, Twitter, January 10, 2019, https://twitter.com/SydneyAzari/status/1083172640947286016.

2. adrienne maree brown, "Pleasure Activism Visuals," *Pleasure Activism* (blog), 2020, https://adriennemareebrown.net/pleasure-activism-visuals/.

3. carla bergman and Nick Montgomery, *Joyful Militancy: Building Thriving Resistance in Toxic Times* (AK Press, 2017).

4. Peter Elbow, "The Believing Game or Methodological Believing," *The Journal of the Assembly for Expanded Perspectives on Learning* 14, no. 1 (January 1, 2008), https://trace.tennessee.edu/jaepl/vol14/iss1/3.

5. Martin Gilens and Benjamin I. Page, "Testing Theories of American Politics: Elites, Interest Groups, and Average Citizens," *Perspectives on Politics* 12, no. 3 (September 2014): 564–81, https://doi.org/10.1017/S1537592714001595.

6. Madison Hopkins and Better Government Association, "As Tons of Chicago Recycling Go to Dumps, a Private Firm Is Paid Twice," Better Government Association, October 10, 2018, https://projects.bettergov.org/2018/recycling-chicago/.

7. Hope Yen, "Washington DC Is Making Public Busses Free Forever," Fortune, December 12, 2022, sec. Politics, https://fortune.com/2022/12/12/washington-dc-making-public-busses-free-forever/.

8. Maylin Tu, "As Fare-Free Transit Catches On, Checking in on 5 Cities With Free Public Transit," NextCity, September 21, 2022, https://nextcity.org/urbanist-news/as-fare-free-transit-catches-on-checking-in-on-5-cities-with-free-public-tr.

9. *Decarb My State*, 2023, https://www.youtube.com/watch?v=cjTKAoF8Cuw.

10. Shari Spiegel and Oliver Schwank, "Bridging the 'Great Finance Divide' in Developing Countries," Brookings (blog), June 8, 2022, https://www.brookings.edu/blog/future-development/2022/06/08/bridging-the-great-finance-divide-in-developing-countries/.

11. Black Maternal Health Caucus, "Black Maternal Health Momnibus," Black Maternal Health Caucus, March 7, 2020, https://blackmaternalhealthcaucus-underwood.house.gov/Momnibus.

CONCLUSION

1. *Armageddon—Animal Crackers (Full HD)*, 2011, https://www.youtube.com/watch?v=cEnWGYagMno.

Bibliography

Aassve, Arnstein, Nicolò Cavalli, Letizia Mencarini, Samuel Plach, and Seth Sanders. 2021. "Early Assessment of the Relationship between the COVID-19 Pandemic and Births in High-Income Countries." *Proceedings of the National Academy of Sciences* 118(36):e2105709118. doi: 10.1073/pnas.2105709118.

AAUW. 2021. *The Simple Truth About the Gender Pay Gap: AAUW Report.* American Association of University Women.

Abdel Salam, Dalia. 2017. "Economic Inequality, Violence, and Life in a Changing Climate—10th World Conference of Science Journalists, San Francisco 2017." Retrieved November 29, 2022 (http://wcsj2017.org/session/climate-data-conflict/).

Ahsan, Sanah. 2022. "I'm a Psychologist – and I Believe We've Been Told Devastating Lies about Mental Health." *The Guardian*, September 6.

Amadeo, Kimberly. 2022. "Living Wage: How Much Do You Need to Live in America?" *The Balance.* Retrieved December 13, 2022 (https://www.thebalancemoney.com/living-wage-3305771).

Anon. 2022c. *The Climate Baby Dilemma.* 90th Parallel Film and Television Productions.

Anon. n.d.-c. "Impacts of the Fossil Fuel Divestment Movemen - Noam Bergman." *Centre on Innovation and Energy Demand (CIED).* Retrieved February 28, 2023 (http://www.cied.ac.uk/publication/impacts-divestment-movement/).

Anon. n.d.-d. "Pluralistic Ignorance: Definition & Examples." Retrieved November 5, 2022 (https://www.simplypsychology.org/pluralistic-ignorance.html).

Arias, Elizabeth, and Farida B. Ahmad. 2022. "Provisional Life Expectancy Estimates for 2021." *Vital Statistics Rapid Release* 16.

Austin, Roy L., and Young S. Kim. 2000. "A Cross-National Examination of the Relationship between Gender Equality and Official Rape Rates." *International Journal of Offender Therapy and Comparative Criminology* 44(2):204–21. doi: 10.1177/0306624X00442006.

Bailey, Stephanie. 2019. "BirthStrike: The People Refusing to Have Kids, Because of 'the Ecological Crisis.'" CNN. Retrieved March 4, 2023 (https://www.cnn.com/2019/06/05/health/birthstrike-climate-change-scn-intl/index.html).

Barros, Beatriz, and Richard Wilk. 2021. "The Outsized Carbon Footprints of the Super-Rich." *Sustainability: Science, Practice and Policy* 17(1):316–22. doi: 10.1080/15487733.2021.1949847.

Barry, Ellen. 2020. "A Massachusetts City Decides to Recognize Polyamorous Relationships." *New York Times*, July 2.

Bascom, Emma. 2022. "Extreme Heat May Increase Miscarriage Risk, Study Shows." Retrieved December 7, 2022 (https://www.healio.com/news/primary-care/20220711/extreme-heat-may-increase-miscarriage-risk-study-shows).

Bashir, Nadia Y., Penelope Lockwood, Alison L. Chasteen, Daniel Nadolny, and Indra Noyes. 2013. "The Ironic Impact of Activists: Negative Stereotypes Reduce Social Change Influence." *European Journal of Social Psychology* 43(7):614–26. doi: 10.1002/ejsp.1983.

Basu, Rupa, Brian Malig, and Bart Ostro. 2010. "High Ambient Temperature and the Risk of Preterm Delivery." *American Journal of Epidemiology* 172(10):1108–17. doi: 10.1093/aje/kwq170.

Basu, Rupa, Varada Sarovar, and Brian J. Malig. 2016. "Association Between High Ambient Temperature and Risk of Stillbirth in California." *American Journal of Epidemiology* 183(10):894–901. doi: 10.1093/aje/kwv295.

Bellucci, Mabel. 1999. "Childless Motherhood: Interview with Nora Cortiñas, a Mother of the Plaza de Mayo, Argentina." *Reproductive Health Matters* 7(13):83–88. doi: 10.1016/S0968-8080(99)90116-7.

Bergman, Carla, and Nick Montgomery. 2017. *Joyful Militancy: Building Thriving Resistance in Toxic Times.* AK Press.

Bernstein, Aaron S., Shengzhi Sun, Kate R. Weinberger, Keith R. Spangler, Perry E. Sheffield, and Gregory A. Wellenius. 2022. "Warm Season and Emergency Department Visits to U.S. Children's Hospitals." *Environmental Health Perspectives* 130(1):017001. doi: 10.1289/EHP8083.

Bidwell, Shelford. 1973. *Modern Warfare: A Study of Men, Weapons and Theories.* Allen Lane.

Black Maternal Health Caucus. 2020. "Black Maternal Health Momnibus." *Black Maternal Health Caucus.* Retrieved December 28, 2022 (https://blackmaternalhealthcaucus-underwood.house.gov/Momnibus).

Brandon, Emily. 2020. "Why Older Citizens Are More Likely to Vote." *US News & World Report.*

Broman, Barbara. 2021. "National Center on Family Homelessness." American Institutes for Research. Retrieved December 30, 2022 (https://www.air.org/centers/national-center-family-homelessness).

Brooks, David. 2020. "The Nuclear Family Was a Mistake." *The Atlantic*, February 10.

Brough, Aaron R., James E. B. Wilkie, Jingjing Ma, Mathew S. Isaac, and David Gal. 2016. "Is Eco-Friendly Unmanly? The Green-Feminine Stereotype and Its Effect on Sustainable Consumption." *Journal of Consumer Research* 43(4):567–82. doi: 10.1093/jcr/ucw044.

brown, adrienne maree. 2020. "Pleasure Activism Visuals." *Pleasure Activism*. Retrieved November 29, 2022 (https://adriennemareebrown.net/pleasure-activism-visuals/).

Brownmiller, Susan. 1976. *Against Our Will: Men, Women, and Rape.* English Language edition. Bantam Books.

Bureau of Labor Statistics. 2021. "Number of Jobs, Labor Market Experience, Martial Status, and Health: Results from a National Longitudinal Survey." Washington, DC: Bureau of Labor Statistics.

Burke, Caro Claire. 2022. "They Took Away Your Rights Because They Knew Your Husband Wouldn't Care." *The Cover Stories.* Retrieved January 2, 2023 (https://caroclaireburke.substack.com/p/they-took-away-your-rights-because).

Burke, Marshall, Felipe González, Patrick Baylis, Sam Heft-Neal, Ceren Baysan, Sanjay Basu, and Solomon Hsiang. 2018. "Higher Temperatures Increase Suicide Rates in the United States and Mexico." *Nature Climate Change* 8(8):723–29. doi: 10.1038/s41558-018-0222-x.

Burke, Marshall, Solomon M. Hsiang, and Edward Miguel. 2014. "Climate and Conflict."

Burke, Susie E. L., Ann V. Sanson, and Judith Van Hoorn. 2018. "The Psychological Effects of Climate Change on Children." *Current Psychiatry Reports* 20(5):35. doi: 10.1007/s11920-018-0896-9.

Bustillo, Miguel, and Kenneth R. Weiss. 2004. "Election Becomes a Fight Over Sierra Club's Future." *Los Angeles Times.* Retrieved December 28, 2022 (https://www.latimes.com/archives/la-xpm-2004-jan-18-me-sierra18-story.html).

Carr, Danielle. 2022. "Mental Health Is Political." *New York Times*, September 20.

Cecco, Leyland. 2018. "Human Rights Groups Call on Canada to End Coerced Sterilization of Indigenous Women." *The Guardian*, November 18.

Cechini, Hannah. 2022b. "Push for Fracking Ban in Delaware River Watershed Continues." 47abc.

Center for Biological Diversity. 2021. "Food Waste Is Trashing the Planet." *Biological Diversity.* Retrieved December 30, 2022 (http://www.takeextinctionoffyourplate.com/waste/index.html).

Centers for Disease Control and Prevention. 2022. "Infant Mortality | Maternal and Infant Health | Reproductive Health." Retrieved December

30, 2022 (https://www.cdc.gov/reproductivehealth/maternalinfanthealth/infantmortality.htm).

Chappell, Bill. 2019. "U.S. Births Fell to a 32-Year Low in 2018; CDC Says Birthrate Is in Record Slump." NPR, May 15.

Chenoweth, Erica. 2013. *The Success of Nonviolent Civil Resistance: Erica Chenoweth at TEDxBoulder.*

Chenoweth, Erica, and Maria J. Stephan. 2011. *Why Civil Resistance Works: The Strategic Logic of Nonviolent Conflict.* Columbia University Press.

Clifford, Cassandra. 2008. "Rape as a Weapon of War and It's Long-Term Effects on Victims and Society." 7th Global Conference: Violence and the Contexts of Hostility, Budapest, Hungary, 13.

Cohn, D'Vera, Juliana Menasche Horowitz, Rachel Mikin, Richard Fry, and Kiley Hurst. 2022. "Financial Issues Top the List of Reasons U.S. Adults Live in Multigenerational Homes." Pew Research Center's Social & Demographic Trends Project. Retrieved November 30, 2022 (https://www.pewresearch.org/social-trends/2022/03/24/financial-issues-top-the-list-of-reasons-u-s-adults-live-in-multigenerational-homes/).

Cohn, Jonathan. 2022. "What the Weird Conservative Fixation On Electric Vehicles Says About Their Politics." *Huffington Post*, October 7.

Collins, Ben. 2015. "Meet the Man Who Beat the KKK with a Tuba." *The Daily Beast*, July 22.

Cong, Zhen, and Merril Silverstein. 2008. "Intergenerational Support and Depression Among Elders in Rural China: Do Daughters-in-Law Matter?" *Journal of Marriage and Family* 70(3):599–612. doi: 10.1111/j.1741-3737.2008.00508.x.

Demby, Gene. 2018. "Our Homeland Is Each Other." NPR.Org.

Dylan, Bob. 2022. "Masters of War." Bob Dylan Official Website. Retrieved February 25, 2023 (https://www.bobdylan.com/songs/masters-war-mono/).

Economic Policy Institute. 2021. "Child Care Costs in the United States." Retrieved December 30, 2022 (https://www.epi.org/child-care-costs-in-the-united-states/).

Ehrlich, Paul. 2018. "Paul Ehrlich: 'Collapse of Civilisation Is a near Certainty within Decades.'" *The Guardian*, March 22.

Elbow, Peter. 2008. "The Believing Game or Methodological Believing." *The Journal of the Assembly for Expanded Perspectives on Learning* 14(1).

Elise, Adams. 2018. "Voluntary Sterilization of Inmates for Reduced Prison Sentences." *Duke Journal of Gender Law and Policy* 26.

Fischer, Dan. 2016. "Bridgeport Residents Release Balloon Banner at City Hall: 'Fracked Gas Is Environmental Racism.'" *Capitalism Vs. The Climate.* Retrieved October 26, 2022 (https://ecology.iww.org/node/1492).

Flory, Richard, and Donald E. Miller. 2016. *The Embodied Spirituality of the Post-Boomer Generations.* Routledge.

Foytlin, Cherri. n.d.-c. "Failing to Save the Earth."

Freedman, Andrew. 2016. "Zika Cases Prompt CDC to Issue a 1-Mile No-Go Zone for Pregnant Women in Miami." *Mashable.* Retrieved December 17, 2022 (https://mashable.com/article/zika-outbreak-miami-cdc-travel -advisory).

Friedan, Betty. 2001. *The Feminine Mystique.* New York: Norton.

Fritz, Angela. 2021. "April Was Earth's 400th Warmer-than-Normal Month in a Row." *Washington Post*, December 3.

Fry, Richard. 2022. "Young Adults in U.S. Are Much More Likely than 50 Years Ago to Be Living in a Multigenerational Household." Pew Research Center. Retrieved March 22, 2023 (https://www.pewresearch.org/fact-tank /2022/07/20/young-adults-in-u-s-are-much-more-likely-than-50-years -ago-to-be-living-in-a-multigenerational-household/).

Gamble, Molly. 2022. "34 States Where Child Care Costs More than College Tuition." *Becker's Hospital Review*, February 17.

Gilens, Martin, and Benjamin I. Page. 2014. "Testing Theories of American Politics: Elites, Interest Groups, and Average Citizens." *Perspectives on Politics* 12(03):564–81. doi: 10.1017/S1537592714001595.

Golden, Gayle. 2017. "Climate Change May Cut the Number of Grandparents." *Next Avenue*, July 12.

Gould, Roger V. 1991. "Multiple Networks and Mobilization in the Paris Commune, 1871." *American Sociological Review* 56(6):716–29. doi: 10.2307/2096251.

Greefux. 2011. *Armageddon - Animal Crackers (Full HD).* https://www.youtube .com/watch?v=cEnWGYagMno

Guthman, Julie. 2011. *Weighing In: Obesity, Food Justice, and the Limits of Capitalism.* University of California Press.

Guttmacher Institute. 2008. "Abortion and Women of Color: The Bigger Picture." Guttmacher Institute. Retrieved December 28, 2022 (https: //www.guttmacher.org/gpr/2008/08/abortion-and-women-color-bigger -picture).

Hamberg, Katarina. 2008. "Gender Bias in Medicine." *Women's Health* 4(3):237–43. doi: 10.2217/17455057.4.3.237.

Hamilton, Brady E., Joyce A. Martin, Michelle J. K. Osterman, Anne K. Driscoll, and Lauren M. Rossen. 2016. *Vital Statistics Rapid Release.* Division of Vital Statistics, National Center for Health Statistics.

Hartmann, Betsy. 1997. "Population Control I: Birth of an Ideology." *International Journal of Health Services* 27(3): 523–40.

——. 2016. *Reproductive Rights and Wrongs: The Global Politics of Population Control.* 3rd edition. Haymarket Books.

Harvey, David. 2007. *A Brief History of Neoliberalism.* Oxford University Press.

Harville, Emily W., Xu Xiong, Gabriella Pridjian, Karen Elkind-Hirsch, and Pierre Buekens. 2009. "Postpartum Mental Health after Hurricane

Katrina: A Cohort Study." *BMC Pregnancy and Childbirth* 9(1):21. doi: 10.1186/1471-2393-9-21.

Haskell, Rob. 2018. "Serena Williams on Motherhood, Marriage, and Making Her Comeback." *Vogue*, January 10.

Hawkins, Derek. 2021. "Judge to Inmates: Get Sterilized and I'll Shave off Jail Time." *Washington Post*, October 25.

He, Chunhua, Cong Liu, Renjie Chen, Xia Meng, Weidong Wang, John Ji, Leni Kang, Juan Liang, Xiaohong Li, Yuxi Liu, Xue Yu, Jun Zhu, Yanping Wang, and Haidong Kan. 2022. "Fine Particulate Matter Air Pollution and Under-5 Children Mortality in China: A National Time-Stratified Case-Crossover Study." *Environment International* 159:107022. doi: 10.1016/j.envint.2021.107022.

Herman, James S. E., J. L. Rankin, M. Keisling, L. Mottet, and M. Anafi. 2016. *Report of the 2015 US Transgender Survey*. Washington, DC: National Center for Transgender Equality.

Hersey, Tricia. 2022. *Rest Is Resistance: A Manifesto*. Little, Brown.

Hill, Elaine L., and Lala Ma. 2022. "Drinking Water, Fracking, and Infant Health." *Journal of Health Economics* 82:102595. doi: 10.1016/j.jhealeco.2022.102595.

Hipple, Kathy. 2019. *IEEFA Update: Fiduciary Duty and Fossil Fuel Divestment*.

Hopkins, Madison, and Better Government Association. 2018. "As Tons of Chicago Recycling Go to Dumps, a Private Firm Is Paid Twice." Better Government Association. Retrieved February 24, 2023 (https://projects .bettergov.org/2018/recycling-chicago/).

Hu, Tianchao, Brett C. Singer, and Jennifer M. Logue. 2012. *Compilation of Published PM2.5 Emission Rates for Cooking, Candles and Incense for Use in Modeling of Exposures in Residences*. LBNL--5890E, 1172959. doi: 10.2172/1172959.

Hurtes, Sarah. 2018. "Women Who Risk Everything to Defend the Environment." *Harper's Bazaar*, November 30.

Jackson, Gabrielle. 2019. "The Female Problem: How Male Bias in Medical Trials Ruined Women's Health." *The Guardian*, November 13.

James, William. 1985. *The Varieties of Religious Experience:* Cambridge, MA: Harvard University Press.

Johannesson, Jessica Gaitán. 2023. *The Nerves and Their Endings: Essays on Crisis and Response*. Scribe US.

Jolly, Shruti, Kent A. Griffith, Rochelle DeCastro, Abigail Stewart, Peter Ubel, and Reshma Jagsi. 2014. "Gender Differences in Time Spent on Parenting and Domestic Responsibilities by High-Achieving Young Physician-Researchers." *Annals of Internal Medicine* 160(5):344–53. doi: 10.7326/M13-0974.

Kallet, Cindy, Ellen Epstein, Michael Cicone, and Richard Knisely. 1993. *Farthest Field*. Overall Music OM-3.

King, Rev. Dr. Martin Luther. 1963. "Letter from a Birmingham Jail."

Kuile, Casper ter. 2020. *The Power of Ritual: Turning Everyday Activities into Soulful Practices*. New York: HarperOne.

Kuile, Casper ter, and Angie Thurston. 2016. *How We Gather*. Cambridge, MA: Harvard Divinity School.

kyodergrist. 2020. "Greentrolling: A 'Maniacal Plan' to Bring down Big Oil." *Grist*. Retrieved March 16, 2023 (https://grist.org/energy/greentrolling-a -maniacal-plan-to-bring-down-big-oil/).

Lalonde, Dianne. 2018. "Sexist Barriers Block Women's Choice to Be Sterilized." *The Conversation*.

Lamidi, Esther O. 2020. "Trends in Self-Rated Health by Union Status and Education, 2000–2018." *SSM - Population Health* 11:100616. doi: 10.1016/j.ssmph.2020.100616.

Laura says every covid death is a policy failure [@TutusNTinyHats]. 2019. "This Is Especially Tricky for Fat People, Because Most of the 'Sustainable,' 'Eco-Friendly,' Etc Brands Don't Make Plus Sizes." *Twitter*. Retrieved October 26, 2022 (https://twitter.com/TutusNTinyHats/status /1198630825312604161).

Lee, Haena, Lindsay H. Ryan, Mary Beth Ofstedal, and Jacqui Smith. 2021. "Multigenerational Households During Childhood and Trajectories of Cognitive Functioning Among U.S. Older Adults." *The Journals of Gerontology: Series B* 76(6):1161–72. doi: 10.1093/geronb/gbaa165.

Leiser, Claire L., Heidi A. Hanson, Kara Sawyer, Jacob Steenblik, Ragheed Al-Dulaimi, Troy Madsen, Karen Gibbins, James M. Hotaling, Yetunde Oluseye Ibrahim, James A. VanDerslice, and Matthew Fuller. 2019. "Acute Effects of Air Pollutants on Spontaneous Pregnancy Loss: A Case-Crossover Study." *Fertility and Sterility* 111(2):341–47. doi: 10.1016/j.fertnstert.2018.10.028.

Levine, Hagai, Niels Jørgensen, Anderson Martino-Andrade, Jaime Mendiola, Dan Weksler-Derri, Irina Mindlis, Rachel Pinotti, and Shanna H. Swan. 2017. "Temporal Trends in Sperm Count: A Systematic Review and Meta-Regression Analysis." *Human Reproduction Update* 23(6):646–59. doi: 10.1093/humupd/dmx022.

Leyser-Whalen, Ophra, Mahbubur Rahman, and Abbey B. Berenson. 2011. "Natural and Social Disasters: Racial Inequality in Access to Contraceptives After Hurricane Ike." *Journal of Women's Health* 20(12):1861–66. doi: 10.1089/jwh.2010.2613.

Lin, C. Y. Cynthia. 2010. "Instability, Investment, Disasters, and Demography: Natural Disasters and Fertility in Italy (1820–1962) and Japan (1671–1965)." *Population and Environment* 31(4):255–81. doi: 10.1007/ s11111-010-0103-3.

Liu, Hui, Zhenmei Zhang, Seung-won Choi, and Kenneth M. Langa. 2020. "Marital Status and Dementia: Evidence from the Health and Retirement

Study." *The Journals of Gerontology: Series B* 75(8):1783–95. doi: 10.1093/geronb/gbz087.

Malik, Rasheed, and Katie Hamm. 2021. "Mapping America's Child Care Deserts." Center for American Progress. Retrieved December 30, 2022 (https://www.americanprogress.org/article/mapping-americas-child-care-deserts/).

Malin, Stephanie A., and Meghan Elizabeth Kallman. 2022. "Enforcing Hopelessness: Complicity, Dependence, and Organizing in Frontline Oil and Gas Communities." *Social Problems* spac032. doi: 10.1093/socpro/spac032.

March of Dimes. 2022. 2022 March of Dimes Report Card.

Margolin, Jamie. 2019. "My Biggest Pet Peeve." Retrieved March 22, 2023 (https://www.facebook.com/watch/?v=353492348781033&ref=sharing).

Marks, Elizabeth, Caroline Hickman, Panu Pihkala, Susan Clayton, Eric R. Lewandowski, Elouise E. Mayall, Britt Wray, Catriona Mellor, and Lise van Susteren. 2021. "Young People's Voices on Climate Anxiety, Government Betrayal and Moral Injury: A Global Phenomenon."

Martin, Bianca. 2020. "What Chicago Can Learn from the 1995 Heat Wave." NPR.org. https://www.npr.org/local/309/2020/07/14/890758229/what-chicago-can-learn-from-the-1995-heat-wave.

Martin, Gina, Kristen Reilly, Haley Everitt, and Jason A. Gilliland. 2022. "Review: The Impact of Climate Change Awareness on Children's Mental Well-Being and Negative Emotions—A Scoping Review." *Child and Adolescent Mental Health* 27(1):59–72. doi: 10.1111/camh.12525.

Matthew. 2015. "History of Mother's Day as a Day of Peace: Julia Ward Howe." The Peace Alliance. Retrieved March 8, 2023 (https://peacealliance.org/history-of-mothers-day-as-a-day-of-peace-julia-ward-howe/).

McAdam, Doug. 1982. *Political Process and the Development of Black Insurgency, 1930–1970*. University of Chicago Press.

———. 1986. "Recruitment to High-Risk Activism: The Case of Freedom Summer." *American Journal of Sociology* 92(1):64–90.

McCormack, Olivia. 2022. "Transgender Advocates Say the End of Roe Would Have Dire Consequences." *Washington Post*, May 7.

McDermott-Levy, Ruth, Nina Kaktins, and Barbara Sattler. 2013. "Fracking, the Environment, and Health." *AJN: The American Journal of Nursing* 113(6):45. doi: 10.1097/01.NAJ.0000431272.83277.f4.

McKibben, Bill. 2012. "Global Warming's Terrifying New Math." *Rolling Stone*. Retrieved December 28, 2022 (https://www.rollingstone.com/politics/politics-news/global-warmings-terrifying-new-math-188550/).

———. 2017. "Stop Talking Right Now about the Threat of Climate Change. It's Here; It's Happening." *The Guardian*, September 11.

———. 2018. "At Last, Divestment Is Hitting the Fossil Fuel Industry Where It Hurts." *The Guardian*, December 16.

McMillan Cottom, Tressie. 2019. "I Was Pregnant and in Crisis. All the Doctors and Nurses Saw Was an Incompetent Black Woman." *Time*.

McMullen, Heather, and Katharine Dow. n.d. "'Too Afraid to Have Kids' – How BirthStrike for Climate Lost Control of Its Political Message." *The Conversation*.

Mendola, Pauline, Maeve Wallace, Beom Seuk Hwang, Danping Liu, Candace Robledo, Tuija Männistö, Rajeshwari Sundaram, Seth Sherman, Qi Ying, and Katherine L. Grantz. 2016. "Preterm Birth and Air Pollution: Critical Windows of Exposure for Women with Asthma." *Journal of Allergy and Clinical Immunology* 138(2):432–440.e5. doi: 10.1016/j.jaci.2015.12.1309.

Mlotshwa, Sikhanyisiwe, Chipo Mutongi, and Conillious Gwatirisa. 2017. "The Impact of the 2015-16 El-Nino Induced Drought on Women." *Journal of Economics and Finance* 2(9):10–17.

Montano, Samantha. 2021. *Disasterology: Dispatches from the Frontlines of the Climate Crisis*. Toronto: Park Row Books.

Morris, Aldon D. 1986. *The Origins of the Civil Rights Movement: Black Communities Organizing for Change*. Simon & Schuster.

mothermade. 2014. "Our Voices." Retrieved November 22, 2022 (https://www.mothermade.us/2014/01/our-voices.html).

Murtaugh, Paul A., and Michael G. Schlax. 2009. "Reproduction and the Carbon Legacies of Individuals." *Global Environmental Change* 19(1):14–20. doi: 10.1016/j.gloenvcha.2008.10.007.

Nagoski, Emily, and Amelia Nagoski. 2020. *Burnout: The Secret to Unlocking the Stress Cycle*. Reprint edition. Random House Publishing Group.

NBC News. "Do You Blast the A/C? Throw out Half Your Lunch? Tell Us Where You Fall Short in Preventing Climate Change." *NBC News*. Retrieved March 14, 2023 (https://www.nbcnews.com/news/specials/climate-confessions-share-solutions-climate-change-n1054791).

Neumayer, Eric, and Thomas Plümper. 2007. "The Gendered Nature of Natural Disasters: The Impact of Catastrophic Events on the Gender Gap in Life Expectancy, 1981–2002." *Annals of the Association of American Geographers* 97(3):551–66. doi: 10.1111/j.1467-8306.2007.00563.x.

Nickalls, Sammy. 2017. "Here's Why All Your Friends Are Deleting Their Uber Accounts." *Esquire*, January 29.

The Onion. 2015. "The Pros and Cons of Fracking." *The Onion*, January 29.

Passy, Florence, and Marco Giugni. 2001. "Social Networks and Individual Perceptions: Explaining Differential Participation in Social Movements." *Sociological Forum* 16(1):123–53.

Paterson, David L., Hugh Wright, and Patrick N. A. Harris. 2018. "Health Risks of Flood Disasters." *Clinical Infectious Diseases* 67(9):1450–54. doi: 10.1093/cid/ciy227.

Perelman, Marcela, and Veronica Torras. 2017. "The Mothers of Plaza de Mayo: 'We Were Born on the March.'" OpenDemocracy. Retrieved December 6, 2022 (https://www.opendemocracy.net/en/mothers-plaza-de-mayo/).

Perry, Imani. 2022. "Why Is 'Auntie' a Controversial Word?" *Unsettled Territory/ The Atlantic*, April 6.

Pew Research Center. 2015. "The American Family Today." Pew Research Center's Social & Demographic Trends Project. Retrieved March 1, 2023 (https://www.pewresearch.org/social-trends/2015/12/17/1-the-american-family-today/).

Poskin, Polly. 2006. *A Brief History of the Anti-Rape Movement*. Illinois Coalition Against Sexual Assault.

Purdue Global. 2023. Generational Differences in the Workplace [Infographic]. Purdue University.

Purdy, Jedediah. 2012. "Occupy Your Voice." *HuffPost*, February 2.

Putnam, Robert D. 2001. *Bowling Alone: The Collapse and Revival of American Community*. First ed. Simon & Schuster.

Ravishanker, Sanchi. 2020. *Children and Climate Change*.

Redden, Molly. 2018. "How America's 'Childcare Deserts' Are Driving Women out of the Workforce." *The Guardian*, January 17.

Reinhart, R. J. 2018. "Global Warming Age Gap: Younger Americans Most Worried." Gallup.Com, May 11.

Relman, Eliza, and Walt Hickey. 2019. "More than a Third of Millennials Share Rep. Alexandria Ocasio-Cortez's Worry about Having Kids While the Threat of Climate Change Looms." *Business Insider*. Retrieved February 12, 2023 (https://www.businessinsider.com/millennials-americans-worry-about-kids-children-climate-change-poll-2019-3).

Rezwana, Nahid, and Rachel Pain. 2021. "Gender-Based Violence before, during, and after Cyclones: Slow Violence and Layered Disasters." *Disasters* 45(4):741–61. doi: 10.1111/disa.12441.

Richter, Roxane, and Thomas Flowers. 2008. "Gendered Dimensions of Disaster Care: Critical Distinctions in Female Psychosocial Needs, Triage, Pain Assessment, and Care." *American Journal of Disaster Medicine* 3(1):31–37. doi: 10.5055/ajdm.2008.0004.

Robinson, Dorothy L. 2015. "Wood Burning Stoves Produce PM2.5 Particles in Amounts Similar to Traffic and Increase Global Warming." *BMJ* 351:h3738. doi: 10.1136/bmj.h3738.

Robles-Fradet, Alexis. 2021. "Medicaid Coverage for Doula Care: 2021 State Roundup." National Health Law Program. Retrieved December 30, 2022 (https://healthlaw.org/medicaid-coverage-for-doula-care-2021-state-roundup/).

Rose, Hilary, and Shannon Hebblethwaite. 2020. "Expanding the Definition of Family to Reflect Our Realities." *The Conversation*. Retrieved October 25,

2022 (http://theconversation.com/expanding-the-definition-of-family-to-reflect-our-realities-131743).

Ross, Loretta, Elena GutiŽrrez, Marlene Gerber, and Jael Silliman. 2016. *Undivided Rights: Women of Color Organizing for Reproductive Justice*. Haymarket Books.

Salamon, Margaret Klein, and Molly Gage. 2020. *Facing the Climate Emergency: How to Transform Yourself with Climate Truth*. New Society Publishers.

Sasser, Jade S. 2018. *On Infertile Ground: Population Control and Women's Rights in the Era of Climate Change*. New York: NYU Press.

Sawicki, Rache. 2022a. "Delaware River Frack Ban Coalition 'Deeply Disturbed' by DRBC Failure to Fully Ban Fracking Activity." *Delaware First Media*.

Schapira, Kate. 2014. "About." *Climateanxietycounseling*. Retrieved December 31, 2022 (https://climateanxietycounseling.wordpress.com/about/).

Schneibel, Andrea. 2018. "Relationship Between Low Income and Obesity Is Relatively New." News, December 11.

Schneider-Mayerson, Matthew. 2018. "The Influence of Climate Fiction: An Empirical Survey of Readers." *Environmental Humanities* 10(2):473–500. doi: 10.1215/22011919-7156848.

Science Daily. "Fracking Industry Wells Associated with Premature Birth." *ScienceDaily*. Retrieved October 26, 2022 (https://www.sciencedaily.com/releases/2015/10/151008110550.htm).

Sharp, Gene. 1973. *198 Methods of Nonviolent Action*. The Albert Einstein Institute.

Shepherd, Laura J., ed. 2014. *Gender Matters in Global Politics: A Feminist Introduction to International Relations*. 2nd edition. London; New York: Routledge.

Shogren, Elizabeth. 2013. "College Divestment Campaigns Creating Passionate Environmentalists." NPR, May 10.

Sierra Club. n.d. "SUSPS - Support US Population Stabilization - Support a Comprehensive Sierra Club Population Policy." Retrieved December 28, 2022 (https://www.susps.org/).

———. "Sierra Club Report Sierra Club Population Report, 1989–SUSPS—Sierra Club Population Immigration Ballot Question." Retrieved December 28, 2022 (https://www.susps.org/history/popreport1989.html).

Solnit, Rebecca. 2016. "'Hope Is an Embrace of the Unknown': Rebecca Solnit on Living in Dark Times." *The Guardian*, July 15.

———. 2020. *Recollections of My Nonexistence: A Memoir*. Illustrated edition. New York: Viking.

St. George, Donna. 2023. "Teen Girls 'Engulfed' in Violence and Trauma, CDC Finds." *Washington Post*, February 13.

Starr, Jared, Craig Nicolson, Michael Ash, Ezra M. Markowitz, and Daniel Moran. 2023. "Assessing U.S. Consumers' Carbon Footprints Reveals

Outsized Impact of the Top 1%." *Ecological Economics* 205:107698. doi: 10.1016/j.ecolecon.2022.107698.

The Stepfamily Foundation. 2022. "Stepfamily Statistics." The Stepfamily Foundation Inc. Retrieved November 29, 2022 (https://www.stepfamily.org/stepfamily-statistics.html).

steve albini [@electricalWSOP]. 2022. "People Are Dunking on These Absurdist Demonstrations but I Think They're Kinda Cool. The Point They're Making Is That Your Most Priceless Shit Is Worthless on a Dead Planet, and While They're Not Going to Change Any Minds It's a Way to Get That Idea into the Discourse." Twitter. Retrieved November 14, 2022 (https://twitter.com/electricalWSOP/status/1580950919940034560).

Sturtevant, Paul B. 2020. "Were the Middle Ages Really 'The Good Place'?" *The Public Medievalist*, January 10.

Supran, Geoffrey, and Naomi Oreskes. 2021. "Rhetoric and Frame Analysis of ExxonMobil's Climate Change Communications." *One Earth* 4(5):696–719. doi: 10.1016/j.oneear.2021.04.014.

Syd [@SydneyAzari]. 2019. "I Want Future Children to Grow up in a World Knowing That the People Alive Today Loved Them So Much They Were Willing to Risk Everything So That They Could Live." Twitter. Retrieved February 28, 2023 (https://twitter.com/SydneyAzari/status/1083172640947286016).

Tapia Granados, José A., Edward L. Ionides, and Óscar Carpintero. 2012. "Climate Change and the World Economy: Short-Run Determinants of Atmospheric CO2." *Environmental Science & Policy* 21:50–62. doi: 10.1016/j.envsci.2012.03.008.

Thomeer, Mieke Beth. 2022. "Relationship Status-Based Health Disparities during the COVID-19 Pandemic." *Social Currents* 23294965221099184. doi: 10.1177/23294965221099185.

Thomeer, Mieke Beth, Jenjira Yahirun, and Alejandra Colón-López. 2020. "How Families Matter for Health Inequality during the COVID-19 Pandemic." *Journal of Family Theory & Review* 12(4):448–63. doi: 10.1111/jftr.12398.

Truven Health Analytics. 2013. The Cost of Having a Baby in the United States. Truven Health Analytics.

Tu, Maylin. 2022. "As Fare-Free Transit Catches on, Checking in on 5 Cities with Free Public Transit." *NextCity*, September 21.

UNHCR, United Nations High Commissioner for. 2016. "Frequently Asked Questions on Climate Change and Disaster Displacement." UNHCR. Retrieved December 13, 2022 (https://www.unhcr.org/news/latest/2016/11/581f52dc4/frequently-asked-questions-climate-change-disaster-displacement.html).

Vetten, Lisa, and Sadiyya Haffejee. 2005. "GANG RAPE: A Study in Inner-City Johannesburg." *South African Crime Quarterly* (12). doi: 10.17159/2413-3108/2005/i12a1017.

Waldinger, Robert J., and Marc S. Schulz. 2010. "What's Love Got to Do With It?: Social Functioning, Perceived Health, and Daily Happiness in Married Octogenarians." *Psychology and Aging* 25(2):422–31. doi: 10.1037/a0019087.

Warren, Team. 2019. "My Plan for Universal Child Care." *Medium.* Retrieved December 30, 2022 (https://medium.com/@teamwarren/my-plan-for-universal-child-care-762535e6c20a).

Webb, Ellen, Julie Moon, Larysa Dyrszka, Brian Rodriguez, Caroline Cox, Heather Patisaul, Sheila Bushkin, and Eric London. 2018. "Neurodevelopmental and Neurological Effects of Chemicals Associated with Unconventional Oil and Natural Gas Operations and Their Potential Effects on Infants and Children." *Reviews on Environmental Health* 33(1):3–29. doi: 10.1515/reveh-2017-0008.

Westervelt, Amy. 2021. "Periodic Reminder That the Window of Opportunity for Moderate Solutions on Climate Was Slammed Shut in Favor of Oil Company Profits for 30 Years and Now All of the Options Are Extreme." *Twitter.* Retrieved March 5, 2023 (https://twitter.com/amywestervelt/status/1439001803878666240).

Whyte, Martin King, Wang Feng, and Yong Cai. 2015. "Challenging Myths About China's One-Child Policy." *The China Journal* 74:144–59. doi: 10.1086/681664.

Win Climate. 2023. *Decarb My State.* https://www.youtube.com/watch?v=cjTKAoF8Cuw

Wynes, Seth, and Kimberly A. Nicholas. 2017. "The Climate Mitigation Gap: Education and Government Recommendations Miss the Most Effective Individual Actions." *Environmental Research Letters* 12(7):074024. doi: 10.1088/1748-9326/aa7541.

Yale Program on Climate Change "Yale Climate Opinion Maps 2021." *Yale Program on Climate Change Communication.* Retrieved October 25, 2022 (https://climatecommunication.yale.edu/visualizations-data/ycom-us/).

Yen, Hope. 2022. "Washington DC Is Making Public Buses Free Forever." *Fortune,* December 12.

Appendix

Beyond Our Bibliography

Here you'll find a list of titles we love, or that are beloved by organizers we interviewed, as well as some newsletters and links to resources that we have found invaluable.

Books

al-Sabouni, Marwa. 2021. *Building for Hope: Towards an Architecture of Belonging.* Thames & Hudson.

Alinsky, Saul D. 1972. *Rules for Radicals: A Pragmatic Primer for Realistic Radicals.* Random House.

Beaton, Kate. 2022. *Ducks: Two Years in the Oil Sands* (graphic novel). Drawn & Quarterly.

Begley, Chris. 2021. *The Next Apocalypse: The Art and Science of Survival.* New York: Basic Books.

Bobo, Kim et al. 2010. *Organizing for Social Change.* The Forum Press.

Boggs, Grace Lee. 1972. *Organization Means Commitment* (pamphlet, freely available online). National Organization for an American Revolution. Retrieved July 4th, 2023 (https://shatteringhegemony.files.wordpress.com/2021/09/organization_means_book_fold_7-8-2012-final.pdf).

Bond, Becky & Zack Exley. *Rules for Revolutionaries: How Big Organizing Can Change Everything.* Chelsea Green Publishing.

Boyd, Andrew. 2023. *I Want a Better Catastrophe: Navigating the Climate Crisis with Grief, Hope, and Gallows Humor.* New Society Publishers.

brown, adrienne maree. 2021. *Holding Change: The Way of Emergent Strategy Facilitation and Mediation.* AK Press.

Foytlin, Cherri. 2022. *The Enchiridion: Failing to Save the Earth, the How-to Guide* (zine). Self-published.

Ghosh, Amitav. 2017. *The Great Derangement: Climate Change and the Unthinkable.* University of Chicago Press.

Johnson, Michelle Cassandra. 2021. *Skill in Action: Radicalizing Your Yoga Practice to Create a Just World.* Shambhala.

Parker, Priya. 2018. *The Art of Gathering: How We Meet and Why It Matters.* Riverhead Books.

Sen, Rinku. 2003. *Stir It Up: Lessons in Community Organizing and Advocacy.* Chardon Press Series.

Schulman, Sarah. 2016. *Conflict Is Not Abuse: Overstating Harm, Community Responsibility, and the Duty of Repair.* Arsenal Pulp Press.

Sherrell, Daniel. 2021. *Warmth: Coming of Age at the End of Our World.* Penguin Books.

Squarzoni, Philippe. 2014. *Climate Changed: A Personal Journey through the Science* (graphic novel). Harry N. Abrams.

Stone, Shoken Michael. 2009. *Yoga for a World Out of Balance: Teachings on Ethics and Social Action.* Shambhala.

Taylor, Richard K. 1977. *Blockade!: A Guide to Non-Violent Intervention.* Orbis Books.

Wray, Britt. 2022. *Generation Dread: Finding Purpose in an Age of Climate Crisis.* Knopf Canada.

Recommended Newsletters

Heated—A newsletter for people who are pissed off about the climate crisis. (https://heated.world/)

Gen Dread—A newsletter about staying sane in the climate and wider ecological crisis. (https://gendread.substack.com/)

Inside Climate News—Weekly digests from one of the largest dedicated environmental newsrooms in the country. (https://insideclimatenews.org/)

Yale Climate Connections—Multimedia stories "connect the dots" between climate change and energy, extreme weather, public health, food and water, jobs and the economy, national security, the creative arts, and religious and moral values, among other themes. (https://yaleclimateconnections.org/sign-up-for-our-weekly-e-newsletter/)

Reasons to Be Cheerful—What it says on the tin. Former Talking Head David Byrne's positive news initiative: "Because the world is full of better ways." (https://reasonstobecheerful.world/)

Low-tech Magazine—In-depth technical articles examine "the potential of past and often forgotten technologies and how they can inform sustainable energy practices. (https://www.lowtechmagazine.com/)

ONLINE RESOURCES

Climable—A woman-led nonprofit in Cambridge, MA, with a mission to make climate science and clean energy understandable and actionable for everyone. (https://climable.org/)

Movement Generation—Movement Generation Justice & Ecology Project inspires and engages in transformative action toward the liberation and restoration of land, labor, and culture. (https://movementgeneration.org/)

The Work That Reconnects Network—The Work that Reconnects helps people discover and experience their innate connections with each other and the self-healing powers of the web of life, transforming despair and overwhelm into inspired, collaborative action. (https://workthatreconnects.org/)

Climate K.I.S.S.—Climate Keep It Simple, Sweetheart: How to talk to kids of different ages to engage them on climate change and environmental issues. (https://climatekiss.com/)

The Ruckus Society—A multi-racial network of trainers dedicated to providing the necessary tools, preparation, and support to build direct action capacity for ecological justice and social change movements. (https://ruckus.org/)

Indivisible—A movement of thousands of group leaders and more than a million members taking regular, iterative, and increasingly complex actions to resist the GOPs agenda, elect local champions, and fight for progressive policies. (https://indivisible.org/)

Rachel's Network—Named for Rachel Carson, they are a granting organization, in their words "a vibrant community of women who catalyze our collective power for a healthy, thriving world." (https://rachelsnetwork.org/)

Decarb My State—Decarb My State allows Americans to picture exactly where their state's carbon pollution comes from, and how to eliminate it. (https://decarbmystate.com/)

B Lab Global Site—B Corp Certification is a designation that a business is meeting high standards of verified performance, accountability, and transparency on factors from employee benefits and charitable giving to supply chain practices and input materials. (https://www.bcorporation.net/en-us/)

Index